T0188011

UNDERSTANDING URBAN METABOLISM

Understanding Urban Metabolism closes the gap between the bio-physical sciences and urban planning and illustrates the advantages of accounting for urban metabolism issues in urban design decisions. *Urban Metabolism* considers a city as a system, and distinguishes between energy and material flows as its components. Based on research from the EU 7th Framework Programme (FP7) Project BRIDGE (sustainaBle uRban plannIng Decision support accountinG for urban mEtabolism), this book deals with the exchanges and transformation of energy, water, carbon and pollutants and introduces a new method for evaluating how planning alternatives can modify the physical flows of urban metabolism components and how environmental and socioeconomic components interact.

The inclusion of sustainability principles into urban planning provides an opportunity to place the new knowledge provided by bio-physical sciences at the centre of the planning process, but there is a strong need for closing the gap between knowledge and practice, as well as a better dissemination of research results and exchange of best practice. This book meets that need and provides the reader with the tools they need to integrate an understanding of urban metabolism into urban planning practice.

Nektarios Chrysoulakis is a Research Director at the Foundation for Research and Technology – Hellas (FOURTH) in Herakleion, Greece. He holds a BSc in Physics, an MSc in Environmental Physics and a PhD in Remote Sensing from the University of Athens. He has been involved in R&D projects funded by organizations such as the European Union, the European Space Agency and the Ministries of Environment, Development, Culture and Education. He has considerable experience in the area of Earth Observation and GIS. His main research interests include urban research, urban energy balance, natural and technological risk analysis, thermal infrared imagery and surface temperature studies, environmental monitoring and change detection. He is the coordinator of the FP7 projects BRIDGE and GEOURBAN. He has more than 100 publications in peer-review journals and conference proceedings.

Eduardo Anselmo de Castro has a degree in Civil Engineering, an MSc in Local Geography, Local and Regional Planning and a PhD in Regional Economics. He is Associate Professor in the University of Aveiro, where he lectures courses on Regional Economics and Planning, and Social and Economic Analysis. Since 1992 he has been a member and regular coordinator of research teams participating in several European and National projects in Innovation and Development Policy, Regional Economics and Regional Policy, Strategic Spatial Planning, Socio-economic Evaluation of Telecommunication Services and Sustainable Development. He is author and co-author of more than 100 papers presented in national and international conferences or published in scientific journals and books.

Eddy J. Moors is Head of the research group Climate Change and Adaptive Land and Water Management of Alterra-Wageningen University and Research Centre, The Netherlands. The group focuses on developing innovative solutions to improve the quality of life within sustainable boundaries. Before coming to Alterra, Moors worked for Wageningen University, the Ministry of Defence of the Netherlands and in Africa and the Caribbean for the World Meteorological Organization. His background is in hydro-meteorology. He has extensive experience in integrating different disciplines to tackle research questions that ask for an inter- and trans-disciplinary approach. He is and has been coordinator of numerous national and international projects ranging from Europe, Africa, India, East-Siberia to the USA and Brazil. He is author and co-author of more than 70 peer reviewed papers and numerous professional publications.

UNDERSTANDING URBAN METABOLISM

A tool for urban planning

Edited by Nektarios Chrysoulakis,
Eduardo Anselmo de Castro and Eddy J. Moors

Routledge
Taylor & Francis Group

LONDON AND NEW YORK

First published 2015
by Routledge
2 Park Square, Milton Park, Abingdon, Oxon OX14 4RN

and by Routledge
52 Vanderbilt Avenue, New York, NY 10017

First issued in paperback 2020

Routledge is an imprint of the Taylor & Francis Group, an informa business

© 2015 selection and editorial material, Nektarios Chrysoulakis, Eduardo Anselmo de Castro and Eddy J. Moors; individual chapters, the contributors

The right of the editor to be identified as the author of the editorial material, and of the authors for their individual chapters, has been asserted in accordance with sections 77 and 78 of the Copyright, Designs and Patents Act 1988.

All rights reserved. No part of this book may be reprinted or reproduced or utilised in any form or by any electronic, mechanical, or other means, now known or hereafter invented, including photocopying and recording, or in any information storage or retrieval system, without permission in writing from the publishers.

Trademark notice: Product or corporate names may be trademarks or registered trademarks, and are used only for identification and explanation without intent to infringe.

British Library Cataloguing-in-Publication Data
A catalogue record for this book is available from the British Library

Library of Congress Cataloging-in-Publication Data
Understanding urban metabolism : a tool for urban planning / edited by Nektarios Chrysoulakis, Eduardo Anselmo de Castro and Eddy J. Moors.
pages cm
Includes bibliographical references and index.
1. Sustainable urban development. 2. City planning—Environmental aspects. 3. Human ecology. I. Chrysoulakis, Nektarios. II. Castro, Eduardo Anselmo de. III. Moors, Eddy J.
HT241.U58 2015
307.1′216—dc23   2014006866

ISBN 13: 978-0-367-67011-5 (pbk)
ISBN 13: 978-0-415-83511-4 (hbk)

Typeset in Bembo
by Swales & Willis Ltd, Exeter, Devon, UK

CONTENTS

FIGURES, TABLES AND PLATES

Figures

Tables

Plates

CONTRIBUTORS

Jorge Humberto Amorim University of Aveiro, Centre for Environmental and Marine Studies, Campus Universitário de Santiago, 3810-193 Aveiro, Portugal

Veronica Bellucco Department of Science for Nature and Environmental Resources, University of Sassari, Via Enrico De Nicola 9, 07100 Sassari, Italy

Carlos Borrego University of Aveiro, Centre for Environmental and Marine Studies, Campus Universitário de Santiago, 3810-193 Aveiro, Portugal

Margaretha Breil Centro Euro-Mediterraneo per i Cambiamenti Climatici S.c.a.r.l., Isola di S.Giorgio Maggiore - 30124, Venezia, Italy, Tel. +39 0412700447, Email: margaretha.breil@feem.it

Acicenta Bubak Institute for Ecology of Industrial Areas Kossutha St., 41-844 Katowice, Poland

Constantinos Cartalis University of Athens, Department of Physics, Panepistimioupolis, Build. Phys–5, 15784, Athens, Greece, Tel. +30 210 72726843, Email: ckartali@phys.uoa.gr

Pedro Cascao Cesam & Department of Environment and Planning, University of Aveiro, 3810-193 Aveiro, Portugal

Eduardo Castro University of Aveiro, Department of Social, Political and Territorial Sciences, Campus Universitário de Santiago, 3810-193 Aveiro, Portugal, Tel. +351 234 370005, Email: ecastro@ua.pt

Nektarios Chrysoulakis Foundation for Research and Technology - Hellas, Institute of Applied and Computational Mathematics, N. Plastira 100, Vassilika Vouton, 70013, Heraklion, Greece, Tel. +30 2810 391762, Email: zedd2@iacm.forth.gr

Manolis Diamantakis Foundation for Research and Technology - Hellas, Institute of Applied and Computational Mathematics, N. Plastira 100, Vassilika Vouton, 70013, Heraklion, Greece

Alison Donnelly Discipline of Botany, School of Natural Sciences, Trinity College Dublin, Ireland

Matthias Falk University of California, Davis, USA Euro-Mediterranean Center on Climate Change (CMCC), Italy

Peter H. Freer-Smith Centre for Biological Sciences, University of Southampton, Building 85 Highfield Campus, Southampton, SO17 1BJ, UK

Ainhoa Gonzáles Discipline of Botany, School of Natural Sciences, Trinity College Dublin, Ireland, Email: agonzal@tcd.ie

C.S.B. Grimmond Department of Meteorology, University of Reading, Reading RG6 6BB, UK, Tel. +44 118 378 6248, Email: c.s.grimmond@reading.ac.uk

Annemarie Groot Alterra - Wageningen University and Research Centre, Wageningen, The Netherlands

Nick Hodges Newcastle University, Rohan, Main Street, Willoughby, Rugby CV23 8BH, UK, Tel. +44 1788 890791, Email: nickhodges281@btinternet.com

Leena Järvi University of Helsinki, Department of Physics, Pl 48, Fin-00014, Helsinki, Finland, Tel. +358 50 3110371, Email: leena.jarvi@helsinki.fi

Mike Jones Discipline of Botany, School of Natural Sciences, Trinity College Dublin, Ireland

Alexandros Karvounis South East European Research Centre, 24 Proxenou Koromila Street, 54622, Thessaloniki, Greece, Tel. +30 2132 038341, Email: akarvounis@seerc.org

Judith E. M. Klostermann Alterra - Wageningen University and Research Centre, Wageningen, The Netherlands, Email: judith.klostermann@wur.nl

Simone Kotthaus Department of Meteorology, University of Reading, Reading RG6 6BB, UK

Björn Lietzke University of Basel, MCR Lab at the Department of Environmental Sciences, Klingelbergstrasse 27, Ch-4056, Basel, Switzerland

Fredrik Lindberg University of Göteborg, Gothenburg, Sweden, Tel. +46 31 7862606, Email: fredrikl@gvc.gu.se King's College London, London, UK

Myriam Lopes University of Aveiro, Centre for Environmental and Marine Studies, Campus Universitário de Santiago, 3810-193 Aveiro, Portugal, Tel. +351 234 372594, Email: myr@ua.pt

Thomas Loridan RMS, London, UK King's College London, London, UK

Vicenzo Magliulo Institute for Mediterranean Agricultural and Forest Systems, Cnr-Isafom, Via Patacca, 85, Ercolano (Napoli) – 80040, Italy, Tel. +39 0817717325, Email: enzo.magliulo@cnr.it

Marta Marques University of Aveiro, Department of Social, Political and Territorial Sciences, Campus Universitário de Santiago, 3810-193 Aveiro, Portugal, Tel. +351 234 370005, Email: marta.marques@ua.pt

Serena Marras Department of Science for Nature and Environmental Resources, University of Sassari, Via Enrico de Nicola 9, 07100 Sassari, Italy. Euro-Mediterranean Centre on Climate Change (CMCC), Via Enrico de Nicola 9, 07100 Sassari, Italy

Helena Martins Cesam & Department of Environment and Planning, University of Aveiro, 3810-193 Aveiro, Portugal

Ana Isabel Miranda Cesam & Department of Environment and Planning, University of Aveiro, 3810-193 Aveiro, Portugal

Zina Mitraka Foundation for Research and Technology - Hellas, Institute of Applied and Computational Mathematics, N. Plastira 100, Vassilika Vouton, 70013, Heraklion, Greece, Tel. +30 2810 391762, Email: mitraka@iacm.forth.gr

Eddy J. Moors Alterra - Wageningen University and Research Centre, P.O. Box 47, 6700 AA, Wageningen, The Netherlands, Tel. +31 317 486431, Email: eddy.moors@wur.nl

Juan Luis Pérez Technical University of Madrid, Environmental Software and Modelling Group, Campus de Montegancedo, Boadilla del Monte, 28660, Madrid, Spain

Gregoire Pigeon Meteo France, Toulouse, France

David R. Pyles University of California, Davis, USA Euro-Mediterranean Center on Climate Change (CMCC), Italy

Roberto San José Technical University of Madrid, Environmental Software and Modelling Group, Campus de Montegancedo, Boadilla del Monte, 28660, Madrid, Spain, Tel. +34 91 336 7465, Email: roberto@fi.upm.es

Mattheos Santamouris National and Kapodistrian University of Athens, Physics Department, Building Physics-5, University Campus, Athens - 15784, Greece

Heikki Setälä University of Helsinki, Department of Physics, Pl 48, Fin-00014, Helsinki, Finland

Donatella Spano Department of Science for Nature and Environmental Resources, University of Sassari, Via Enrico de Nicola 9, 07100 Sassari, Italy, Tel. +39 079 229339, Email: spano@uniss.it. Euro-Mediterranean Centre on Climate Change (CMCC), Via Enrico de Nicola 9, 07100 Sassari, Italy

Tomasz Staszewski Institute for Ecology of Industrial Areas Kossutha St., 41-844 Katowice, Poland

Marina Stathopoulou University of Athens, Department of Physics, Panepistimioupolis, Build. Phys-5, 15784, Athens, Greece

Afroditi Synnefa National and Kapodistrian University of Athens, Physics Department, Building Physics-5, University Campus, Athens - 15784, Greece

Matthew James Tallis School of Biological Sciences, University of Portsmouth, King Henry Building, King Henry 1 Street, Portsmouth, PO1 2DY, UK. Centre for Biological Sciences, University of Southampton, Building 85 Highfield Campus, Southampton, SO17 1BJ, UK

Richard Tavares Cesam & Department of Environment and Planning, University of Aveiro, 3810-193 Aveiro, Portugal

Pierro Toscano Institute of Biometeorology (IBIMET – CNR), Via G.Caproni 8, 50145 Firenze, Italy

Frank van der Bolt Alterra - Wageningen University and Research Centre P.O. Box 47, 6700 AA, Wageningen, The Netherlands

Ab Veldhuizen Alterra - Wageningen University and Research Centre, P.O. Box 47, 6700 AA, Wageningen, The Netherlands

Roland Vogt University of Basel, MCR Lab at the Department of Environmental Sciences, Klingelbergstrasse 27, Ch-4056, Basel, Switzerland, Tel. +41 61 2670 700, Email: roland.vogt@unibas.ch

Duick T. Young Department of Meteorology, University of Reading, UK

ACKNOWLEDGEMENT

The research leading to the results presented in the book has received funding from the European Community's Seventh Framework Programme (FP7/2007–2013) under grant agreement n° 211345 (the BRIDGE Project: http://www.bridge-fp7.eu). The authors are grateful to all the people who have contributed to the BRIDGE research, as well as to the participants in each of the cities.

PART I
Introduction

1

URBAN METABOLISM

Alexandros Karvounis

SOUTH EAST EUROPEAN RESEARCH CENTRE

Introduction

This chapter deals with the current dynamics of urban systems, introducing the 'urban metabolism' concept and its importance in sustainable urban planning. Due to radical changes in ecology, socio-economic values and environmental quality, long term efficiency of cities is questioned. Urban planning seems to be an effective tool for the necessary reformations towards a sustainable city model, while constraints confronted in urban processes are well described. Community empowerment and flexible legislation are getting involved in the changes needed. The evolution of urban centres has caused a spatial segregation and competition that takes place within cities. At the same time the resource flow systems of the cities seem to move towards saturation (Batty 2008). Cities are no longer only economic stimulators, but also social, cultural and ecological motors towards sustainable development. Environmental problems are nowadays so severe that globally cities face them in everyday life activities. Moreover, consumption is increasing the demand of resources, resulting in complex waste flows that call for sustainable solutions (Rotmans and Van Asselt 2000).

The aspect of life quality gains importance for planning strategies and is now on the core of actions with the establishment of planning systems by city governments. The recognition of the fact that cities are a human creation, contributes to the understanding of the role of nature and interactions with the flow of resources needed with growth (Pincetl 2012). So, urban transformations necessitate city planning to integrate physical, social and cultural infrastructure, the economy and the environment of the city. As an extent, new strategies and mechanisms are the means necessary to promote flexibility in commuting, supply of power, equal water distribution and effective waste management system (Chrysoulakis et al. 2013).

In an effort to deal with the appearance of new large scale problems, research is focused on the efficiency of the city. Sustainability is entering the field and the understanding over the new contents of the city requires to be discussed (Pincetl 2012). Due to previously described problems that take place globally, there is an extensive need to quantify the aspects of consumption and waste production of urbanizing areas. In terms of this large increase in the demand of resources, the concept of urban metabolism has been introduced, which compares a city to an organism, in their common trait of the demand for food and the deposition of waste on the environment (Grimm et al. 2008). The spatial heterogeneity and various local ecosystems reveal that reference to organisms is more appropriate than a comparison with ecosystems as they evince the high reliance of socio-economic factors in the total considering of urban metabolism concept (Golubiewski 2012), especially if someone considers the dynamic character of biological and physical flows that take place.

Nevertheless, opinions differ, as from one point of view, only individual organisms have a metabolism and from the other, cities are presumed more like ecosystems, an aggregation of various metabolisms. As ecologists support, an ecosystem embodies interactions among various individuals. So, we can analogize the city better as an ecosystem than an organism (Pincetl 2012). Further, the fact that a city seems to function as an ecosystem, supports the research to this way (Golubiewski 2012). At this point, it is fundamental to highlight that although natural ecosystems are considered in general terms as energy self-sufficient, urban ecosystems' metabolic cycles seem unsustainable (Chrysoulakis et al. 2013).

As Kennedy et al. (2007) argue 'urban metabolism can be defined as the sum total of the technical and socio-economic processes that occur in cities, resulting in growth, production of energy, and elimination of waste'. More specifically, it describes the processes of resource inputs, the way they are distributed within the system, how they convert to products ready to be consumed and the concern over their recycling (Zhang et al. 2012). At the end of the previous century, via a system based approach, city and flows within it were presented by their impacts on the environment (Pincetl et al. 2012).

The concept of urban metabolism was explored through ecology and industrial ecology, bio-physical sciences, political economy and urban planning studies, revealing in each case different aspects integrated in the concept (Pincetl et al. 2012). For example, urban and industrial ecologists deal with circulation of materials within urban systems while ecological economists from one point of view focus on the interdependent relationship between nature and urban economy; at the same time there are opinions that examine the way urban metabolisms may contribute to the production of inequality globally (Rapoport 2011). A broader concept is described by the urban metabolic system that Zhang et al. (2012) introduce, considering cities as socio-economic systems with both industrial and consumption components (Zhang et al. 2012). Nowadays, urban metabolism becomes more and more topical, as it appears to be a tool that, through understanding of bio-physical systems, enables the management and integration of ecological and socio-economic processes that take place within urban systems (Golubiewski 2012).

The 'urban metabolism' concept

Environmental flows

In 2005, the Millennium Ecosystem Assessment (MEA) emphasized the relation between environmental flows and socio-economic structures with regard to the evolution of urban ecosystems. The MEA estimated the consequences on ecosystems' changes and highlighted the contribution of the human factor. This plight is well understood through the 1987 Brundtland Commission report, with ecosystem health to be its basic axis, drawing attention to environmental deterioration and setting the foundations for raising awareness before an ecological collapse. Furthermore, Agenda 21 for cities, a new agenda, supported Brundtland Commission's assertions by paving the way towards a sustainable city model (Pincetl 2012).

The urban metabolism concept was first introduced by Marx who proposed an urban–rural metabolism, suggesting the transformation of human and animal wastes into fertilizers in order to implement a circular process for urban ecosystems. However, the environmental implications of urbanization are also evident through the 'metabolic rift' concept he introduced, which, additionally, attempted to study social influence. The population increase and the human migration from rural to urbanized areas created a rift in the metabolism processes, since human beings altered the cyclical character of many flows, obstructing the return of wastes to the soil. In the 20th century, Eugene Odum pointed out the biological aspect of the urban metabolism concept (Golubiewski 2012; Rapoport, 2011) and dealt with conceptualization of energy. However, the concept was firstly applied by Wolman in 1965, to a hypothetical city. He comprehended the complexity of urban systems and studied the dynamic socio-ecological flows that should be evaluated (Pincetl 2012).

More specifically, urban metabolism is determined by the quantification of inputs and outputs flows, in order for conclusions on the balances of ecosystem to be drawn (Pincetl 2012). Nutrients, energy, materials, water capacity and wastes are all calculated, while their life cycle is integrated into the concept as well (Pincetl 2012; Wachsmuth 2012). The system that functions within a city relies on a circulatory network where materials and wastes are moving through. Metabolic flows are derived from urban development, food consumption, energy and material use; as a city develops and grows all the human-made circulate systems have to expand and the environment should be ready to accept the increasing disposal of wastes. Nonetheless, the waste management needs to consume energy in order for garbage to be reused or recycled (Golubiewski 2012).

An important impact of globalization, with regard to urban metabolism, is that the circulation of soil nutrients has changed. Due to population demand and globalization, nutrients do not stay at the place they are produced so the cycle of food changes. Nutrients are currently consumed far away from the place of origin (Wachsmuth 2012). The bigger the population demand, the bigger the crop capacity with respect to global market competition rules. Additionally, productivity has been reinforced by the application of fertilizers, and provision of nitrogen, phosphorus and potassium to the soil is intensified (Villarroel-Walker and Beck 2012).

Carbon, in turn, is placed at the top of the list of nutrients that urban metabolism studies have calculated. Its importance is evident, for instance, through the carbon footprint, which nowadays is used as a tool for quantification of sustainability aspects (Villarroel-Walker and Beck 2012). Carbon is primarily connected to greenhouse gases released in the atmosphere, as cities produce carbon dioxide (CO_2) and as a consequence of an imbalance in biogeochemical processes. Urbanization is followed by climate change which is increased by atmospheric carbon concentrations. CO_2 appears to be the end-stage of waste decomposition of all products while its increase is caused by the demand for nutrients and fossil fuel burning. The tendency of cities to use renewable energy is promising a decrease of carbon fluxes dependency.

As for energy matters, urban metabolism studies examine the main types that include fuels, electricity, radiation and heat, although we cannot ignore the stored energy that construction of materials and food and waste management uses (Chrysoulakis et al. 2013). Additionally, it is important to mention that when we refer to energy consumption, we focus on non-renewable forms of energy, which do not promote a sustainable solution. With regard to urban metabolism, energy flows can contribute to understanding how rural and urban areas use them and balance, or not, their supply and demand needs (Villarroel-Walker and Beck 2012). Guidelines can regulate and diminish energy consumption, while citizens themselves can contribute simply by choices of eco-friendly materials (Loridan and Grimmond 2012). It is urgent that urban heat island (UHI) effect is considered in urban planning, as it takes place locally while having a regional/global impact. Causes such as impermeable surfaces, city size, land cover pattern, scarcity of green and open water bodies can drive its intensification. Impacts from inefficient energy fluxes are serious, especially if we consider the possible relation to the increasing flood risks that most European cities experience (Grimm et al. 2008).

Human activities have altered the cycle of water, as urbanization drives forces to expand infrastructure, by constructing reservoirs or artificial water entities. In addition we cannot ignore the expanded dimensions of current pollution events. Fertilizers, air pollutants and acid rain are common reasons that cause water resources deterioration. In cities, mainly in North America, wastewater systems are still not separated from storm water infrastructure (Grimm et al. 2008; Villarroel-Walker and Beck 2012). But the quality of water also decreases as the number of organisms increase because of the intake of food in household waste (Villarroel-Walker and Beck 2012). Within cities the pollution is getting worse due to conflicting land uses, such as the construction of road networks and residences into river beds (Grimm et al. 2008). Additionally, while modern cities should confront pollution, in the developing world the situation is even more desperate with millions of people denied access to potable water (Hallsmith 2007).

Concluding, most present day cities are based on unsustainable cycles that are open and imbalanced between inputs and outputs, with a relation between inputs and outputs that tends to be linear. In contrast, natural systems have a cyclical course of flows. Cities' goals are to concentrate attention on a decrease of inputs, or effective reuse and recycling of waste. Linearity of urban flow systems have to be brought to an end and need to be replaced by a cyclical course.

Socio-economic values

Urban metabolism requires a wider meaning in order to include political and social factors in the study of urban phenomena. Moving beyond the nature–society dualism that incorporates the metabolic process of today's theories, the urban metabolism concept led to a reconceptualization, taking into account social and economic factors. The social factor is to be considered as highly important because urban metabolism refers to peoples' desires and visions, which shape the material and energy flows and also control the waste output of cities. This process embodies the way political and human perspectives enter the game (Rapoport 2011).

Characteristically, the urbanization process requires an understanding of changes in socio-environmental, economic and legislative aspects. In no case, can we analyse the dynamic phenomenon of urbanization based exclusively on the flow of materials (Golubiewski 2012). Social networking with its interconnections and the evolution of residents' location may be linked to the urban living environment (Grimm et al. 2008). Quality of life and prosperity for everyone should be set as a goal and involvement of businesses, non-profit organizations and civil society will enable the implementation.

Human impacts and, thus, social drivers play a basic role in the metabolism activation of each city. Social welfare is based on a sense of community, safety, equity and health care, education, chances, spiritual development and aesthetic life. It is accepted that not all actors can actively contribute to urban metabolism in the same way, as it is subject to current social networks; citizens' needs, in turn, cannot be satisfied to the same degree. Nevertheless, the impacts of urban metabolism processes within a city remain fundamental due to their global effects (Rapoport 2011).

The needs are represented by supply and demand, as the increase of population is testing the city's ability to meet the new needs. Resource consumption and high demand for energy, technological innovations and economies based on networks that promote global interconnectivity are observed in each industrialized urban centre (Rapoport 2011). It is widely accepted that economic forces lead to cities to grow or shrink, an important aspect of urban metabolism, which is moving in parallel to political–economic vectors at all levels of human organization, either local or global (Pincetl et al. 2012). Thus, it is important to evaluate socio-economic metabolism and come up with energy consumption independent/self-sufficient economies and material-independent economies (Villarroel-Walker and Beck 2012).

The urban metabolism concept, as it describes resource exploitation and energy use through the demand of inputs, quantifying environmental flows, can give these high rates of consumption by economic activity (Golubiewski 2012). Material resources that flow within the city trigger economic and social development and their effective use determines how much waste, emissions, effluents and resource shortages they create (Karvounis 2009). The material fluxes can alter depending on economic activities that take place in the area. For example, the high concentration of phosphorus used in Finland is justified by the quantities of manure produced by the livestock industry (Villarroel-Walker and Beck 2012).

Of particular concern is the fact that many believe that economic growth itself is a significant factor that leads to environmental deterioration. As industrial and urban ecology defines, city growth and expansion, energy consumption and waste elimination are all processes that are implemented in cities, and metabolism is defined by integrating all these processes. In particular, during the last centuries, the intensification of urbanization has demonstrated the relationship between economic abundance and prosperity with environmental exhaustion. However, in terms of sustainable urban metabolism, industrial ecologists try to eliminate the resources used per unit of economic output with a process referred to as dematerialization, in order to ameliorate metabolic efficiency (Golubiewski 2012; Rapoport 2011). Further, the link between economic development and environmental decline can be easily explained if we refer to Marx's idea of metabolic rift (Rapoport 2011).

Towards a sustainable city model

Bio-physical sciences' research has to be totally integrated into urban planning at different scales. Even in a neighbourhood or a city, planning has to promote sustainable practices, and transformations taking place have to deliver sustainability. During the last decades, sustainable development and environmental protection came to the foreground, highlighting the necessity of understanding the interrelationship between resource flows and spatial structures within a city (City of Kitchener 2007). Besides, nature influenced city models such as Ebenezer Howard's Garden Cities and Le Corbusier's Contemporary City, while planners as Scott Campbell linked sustainability principles to ecosystem services (Pincetl 2012).

On a more practical level, a sustainable city can be designed with a combination of practices and strategies described below. Urban tissue and built environment can largely succour the efforts towards sustainability upgrade. The goals remain to be the creation of a city in which functions and infrastructure amplify social prosperity, economic equity and environmental balance. This implies that the built environment such as buildings, open and green spaces, roads and different kinds of infrastructure achieve the previously described objectives. Core characteristics for sustainable cities should include walkability, liveability, conservation and safety. More specifically, cities should promote inter-connectivity of areas through walkable tours where citizens feel safe and can 'experience' the city itself. Transit capacity that a city provides is now a primary goal of urban regeneration and aims at the interchange of different modes of transportation (City of Kitchener 2007). In urban planning terms, local authorities can publish regulations that stop urban sprawl by intensifying housing in existing areas, encouraging a compact city model. Similarly, mixed land-use strategies can be proved as efficient enough for the creation of sustainable societies. If decision makers combine residential with commercial land uses at the neighbourhood level, they can boost employment opportunities while the proximity to workplace reduces car congestion and, consequently, the need for energy, noise levels and air pollution. Built environment has to attach a special sense of place to the neighbourhood, respecting a high-quality landscape at the same time. The internal environment of urban centres has to be healthy and active enough in order for citizens to have the opportunity of a great quality of life. Old urban fabric elements can be totally combined to create new materials, for example old stone houses can be regenerated with the addition of steel. These proposals can be easily presented through advanced design software. Design of the built environment can also give urban centres the chance to adapt to the cycle of growth and transform their parts without ruining the whole urban tissue (Kennedy et al. 2011; Rapoport 2011).

As we understand from the bio-physical flows that create the metabolism of a city, transit capacity of a city remains one of the most fundamental goals to achieve. Independent of the size of a city, urban sprawl has to come to an end and density needs to keep going up. Planners have to promote the pedestrianization of many roads, avert the parking of cars and propound the beneficial use of alternative modes of transportation. Open spaces and squares in cities should be well connected through bicycle lanes or pedestrian friendly routes where citizens will be safe to move around. Additionally, a more sustainable urban form will be further supported by updated land use plans and policies which promote a reduction of the travel time within neighbourhoods, created by the coexistence of residential and commercial uses, offices and community facilities. As a consequence, air pollutants per locality can be significantly reduced, eliminating the impacts on the greenhouse and UHI effects by the increase of residents' proximity to public transport hubs. Renewable energies have to invade our lives and energy efficiency policies have to be implemented. In parallel with transportation systems upgrade, greening of the cities and conservation standards for new and existing buildings should be completely adapted and implemented in each city globally; these kinds of strategies can become attractive through the provision of tax incentives. Water pollution and the rapid rate of water extraction require sustainable treatment and planning ahead; therefore a great deal of attention must be given to the waste management sector. Actions to lower demand and consumption appear to be the first strategies that need to be introduced, while at the same time reuse and recycling must replace garbage disposal completely (Karvounis 2009).

Climate change can be mitigated by 'greening the city' actions. Urban reforestation, preservation of green spaces and creation of gardens, on balconies or roof gardens, are the short term actions that can be implemented by the citizens themselves. But the hard part of this task is that, in many cases, strengthening of the sense of community is required (Kennedy et al. 2011). The NIMBY (Not In My Back Yard) syndrome has to cease and be replaced by acting together for a common goal. These kinds of urban strategies can improve local climate conditions, understate the UHI effect and even diminish water needs. Air flow in the city will be ameliorated and CO_2 will be absorbed by the plants. Highly beneficial is the increase of urban agricultural practices in backyards and unused plots with the objective of directly supplying food to consumers. The sense of community will be amplified while environmental health will improve.

Self-sustaining communities are based on more sustainable economies that are supported by individuals who have concern for environmental safeguards (Rapoport 2011). An economic approach which focuses not only on job creation or profit-making, but also on innovation, has to transmute the business base. Green businesses have to be attracted in spatially well-defined areas where zoning is reinforced by financial and tax incentives. Thereby policy makers and local authorities can promote sustainable production and consumption through the spatial concentration of businesses that invest only in environment friendly technologies and form a network that allows the exchange of know-how, the provision of training or even a great accessibility to natural resources located in high proximity (Roseland 2005).

Contributions of urban metabolism to planning and decision making

Policy tools

Urban planning can reform dynamics within city boundaries and small scale interventions in policy making can have results seen globally. According to Pincetl (2012), comprehensive planning can open democratic processes. Planning tools should include community participation and empowerment during decision-making and planning processes. Community empowerment and engagement are essential in order to achieve the goals and objectives determined.

Understanding by policy makers of the urban metabolism concept and how it can contribute to everyday life would be of great benefit. To achieve this benefit, it is essential for urban policy makers to know about risks regarding resource exhaustion (Kennedy et al. 2007). Once the end users acquire the big picture, they can fill the gaps and seek new solutions for interconnection of different sectors in order to improve effectiveness. A sustainable city model could provide a good background for the adoption of new policies and strategies enabling cities to reduce their reliance on inputs from distant places, to decrease waste streams and to achieve social equity (Chrysoulakis et al. 2013). In addition, awareness of the financial, social and environmental needs of local population is fundamental, since by applying systems theory, mechanisms for indication and evaluation could be set up. However, it is important to consider that planning is not only concerned with short term interventions as the long term considerations are more important than ever, due to the current global problems (Hallsmith 2007). Thus, working towards new policy goals is considered one-way, and the search for quality is a search for long term development, overcoming barriers such as the lack of (both quantitative and qualitative) data (Arbor 1999).

Policy and decision-making frameworks can give new chances for experimentation. New practices have to be implemented, risks defined and limitations of current policies as well as increasing financial difficulties overcome. Policy reformation becomes necessary as a response to global problems, which should be addressed first by governance (Pincetl et al. 2012). Further, the policy and legislation framework can embody the urban metabolism concept in the definition of new objectives and guidelines. An environmental approach has to be totally integrated with socio-economic challenges of the current era and, finally, urban policies have to be updated. Furthermore, an interdisciplinary approach can clarify the effect that legislation has on urban flows and can contribute to more efficient management (Villarroel-Walker and Beck 2012).

A difficult task to be achieved continues to be the incorporation of different levels of action, as individuals have to change behaviour. Government and local authorities have to be able to formulate regulations (standard setting) depending on needs, and private agencies have to be able to integrate new technologies such as computational modelling with the means to provide multi-disciplinary approaches as tools for both decision making and future scenarios evaluation (Villarroel-Walker and Beck 2012). The cooperation and coordination of administrations at different scales should promote local plans to be integrated with national policies, resulting in city planning being more integrated. Social organization is essential for a city's function. Thus, human needs act as a catalyst for decision making that shapes both urban environment and local natural ecosystems (Pincetl 2012). Urban governance is a benchmark for organization and social integrity, requiring proactive participation from citizens and motivating them to act within a democratic framework (UNEP 2008). If urban metabolism is integrated in planning policies for the analysis of cities, it requires the examination of each component of the urban cycle, such as inputs and waste flows, and current socio-economic structures, as well as taking into account the legislation and policy drivers that influence these processes (Pincetl et al. 2012).

Policy recommendations have to be economic centred and broaden their effect (Rapoport 2011). The positive and negative effects of a city have to be considered as a reflection of the modification that sectoral policies suggest, even if we refer to the internal metabolism of the system or the system built beyond its boundaries (Villarroel-Walker and Beck 2012). This turnover is fundamental, even though there are still opinions that express that this kind of political relevant strategy does not lead to a complete reformation for effective urban planning interventions (Rapoport 2011). It is worth mentioning the significant proportion of uncertainty that may be associated with the results of the previously described practices. However, even failing practices can conduct and boost research in-depth (Villarroel-Walker and Beck 2012).

Socio-economic incentives

Efficient economic management can add to human and social development and also contribute to sustainable urban progress. In economic terms, sustainability with regard to local and global scales is elevated into a fundamental factor towards social integrity. The intense gap between rich and poor has to be gradually eliminated and the foundations for sustainable development to be set up (Donovan et al. 2005). Economic incentives have to be reformed as have national bodies, and finance organizations, since the policies and the objectives they set up can regulate markets at an international level, and because their influence in economic terms is evident considering the world water market. Individuals have an important role in the implementation of the vision and targeted policies – it is hard to alter current lifestyles and mentality. For this reason, a sound understanding of the daily practices and local political economies is needed for governance, as it facilitates or baulks urban resource flows and networks (Rapoport 2011).

New legislation has to stop supporting environmentally controversial industries and give incentives to businesses and companies that focus on energy efficiency and low waste. Government regulations should decrease externalities that the current market conditions cause and costs have to be a primary target of policy makers. Regulation can be in the form of penalties and fines that will burden citizens and businesses that exceed the environmental standards and limitations a society has set. In particular, regarding sustainability strategies, the impact, either social or economic, must be taken into consideration. Investments for sustainable development have to be supported and reinforced, while taxation and planning regulation need to be downscaled and provide attractive incentives to citizens. Moreover, economic incentives can impose cost and revenue imperatives mainly on businesses. For instance, green energy-based corporations can achieve lower costs and enhanced revenues from having a good reputation (Epstein 2008), although the risk of a 'rebound effect' is still evident. As products become more eco-friendly, opportunities for increasing consumption are fostered, and as a result a coherent market strategy has to be implemented at the same time as sustainability action plans (UNEP 2008).

However, a change in laws and regulations will probably impose a conflict of interest, between consumers and producers for example, and, thus, a fair balance should be achieved. Self-sufficient and zero carbon buildings, eco-friendly transportation and recycling are the pillars of a sustainable strategy which focuses on everyday life standards. Last but not least, decision makers and planners should integrate sustainable technologies in design and construction and provide motives to support this way of working (Karvounis 2009). Regarding material flow analyses, taxing resource consumption or subsidy policies have to be replaced by labour-led solutions and environmental control regulations (Rapoport 2011). Financial incentives are a good way to encourage stakeholders to apply new practices, fiscal policies have to be sensitive to the environment, and eco-taxation or subsidies on green technology have to be imposed.

Conclusions

Urban metabolism studies quantify flows of resources within a city, describe how its circulation system functions and evaluate the impacts on the environment. Decision makers and planners have to collaborate and understand how urban metabolism can be applied practically on urban tissue. Exploration of the metabolism concept showed the importance of the conjunction of physical flows with built environment, as social values call attention towards the shaping of a sustainable city model. Equity, sense of community, provision of infrastructure for health, education and recreation are only a few of the standards that an urban centre has to be able to provide to everyone, without considering their income.

The biggest challenge to face remains the expansion to different scales of decision making. As urban metabolism rules, social, economic, geographical, political and institutional objectives have to be discussed all over again and be defined on a new foundation of standards. The integration of governance structures with economic actors or legislation constraints or updates presents a weak point of the urban metabolism concept; the present theoretical framework has to be further developed.

All the previously described global problems have brought to light the need for top-down policy making that ameliorates conditions at a local scale but has global consequences. A basic disadvantage to highlight for urban metabolism is the current way of thinking. Our societies do not respect nature at all and suffocate inside the boundaries of personal interest. Sense of community has to be reinforced and every kind of policy and strategy needs to promote collective action. Urban metabolism studies begin to function in a positive way for the cities. Sustainability is adopted in development processes and the world learns about its extra-beneficial impacts. Sustainable urban metabolism needs stakeholders to promote participatory planning, as citizens are the basic actors that should have access to every stage of planning, knowing both their own and their cities' real requirements.

References

Arbor, A. (1999). Sustainability obstacles. National Sustainable Buildings Workshop, October 8–9.

Batty, M. (2008). The Size, Scale and Shape of Cities. *Science*, 319, 769–771.

Chrysoulakis, N., Lopes, M., San José, R., Grimmond, C.S.B., Jones, M.B., Magliulo, V., Klostermann, J.E.M., Synnefa, A., Mitraka, Z., Castro, E., González, A., Vogt, R., Vesala, T., Spano, D., Pigeon, G., Freer-Smith, P., Staszewski, T., Hodges, N., Mills, G. & Cartalis, C. (2013). Sustainable Urban Metabolism as a Link between Bio-physical Sciences And Urban Planning: The BRIDGE Project. *Landscape and Urban Planning*, 112, 100–117.

City of Kitchener (2007). Design Brief for Suburban Development and Neighbourhood Mixed Use Centres. Online. Available HTTP: <http://www.kitchener.ca/Files/Item/item11097_dts-07–065_-appendix_j__design_brief_-_april_23_.pdf (accessed March 2009)>.

Donovan, R., Evans, J., Bryson, J., Porter, L. and Hunt, D. (2005). Large-scale Urban Regeneration and Sustainability: Reflections on the "Barriers" Typology, Working Paper 05/01, University of Birmingham.

Epstein, M. (2008). *Making Sustainability Work*, UK: Greenleaf Publishing

Golubiewski, N. (2012). Is There a Metabolism of an Urban Ecosystem? An Ecological Critique. *AMBIO*, 41, 751–764.

Grimm, N., Faeth, S., Golubiewski, N., Redman, C., Wu, J., Bai, X. and Briggs, J. (2008). Global Change and the Ecology of Cities, *Science*, 319, 756760.

Hallsmith, G. (2007). *The Key to Sustainable Cities; Meeting Human Needs, Transforming Community Systems*, Canada: New Society Publishers.

Karvounis, A. (2009). Protocol to Assess Differences between Knowledge Supply and Knowledge Needs in the field. BRIDGE Deliverable D.2.2. Online. Available HTTP: <http://www.bridge-fp7.eu (accessed August 2013)>.

Kennedy, C., Cuddihy, J. and Engel Yan, J. (2007). The Changing Metabolism of Cities, *Journal of Industrial Ecology*, 11, 43–59.

Kennedy, C., Pincetl, S. and Bunje, P. (2011). The Study of Urban Metabolism and its Applications to Urban Planning and Design. *Environmental Pollution*, 159, 1965–1973.

Loridan, T. and Grimmond, C. (2012). Characterization of Energy Flux Partitioning in Urban Environments: Links with Surface Seasonal Properties, *Journal of Applied Meteorology and Climatology*, 51, 219–241.

Pincetl, S. (2012). Nature, Urban Development and Sustainability - What New Elements are Needed for a More Comprehensive Understanding? *Cities*, 29, S32–S37.

Pincetl, S., Bunje, P. and Holmes, T. (2012). An Expanded Urban Metabolism Method: Toward a Systems Approach for Assessing Urban Energy Processes and Causes. *Landscape and Urban Planning*, 107, 193–202.

Rapoport, E. (2011). Interdisciplinary Perspectives on Urban Metabolism, A Review of the literature. UCL Environmental Institute Working Paper, UCL: Development Planning Unit.

Roseland, M. (2005). *Toward Sustainable Communities; Resources for their Citizens and their Governments*, Canada: New Society Publishers.

Rotmans, J. and Van Asselt, M. (2000). Towards an Integrated Approach for Sustainable City Planning. *Journal of Multi-criteria Decision Analysis*, 20, 110–124. Online. Available HTTP: < http://www.icis.unimaas.nl/ (accessed 16 March 2009)>.

UNEP (2008). *Planning for Change. Guidelines for National Programmes on Sustainable Consumption and Production*. United Nations Environment Programme. UK: Waterside Press. Online. Available HTTP: < http://www. unep.org/pdf/UNEP_ Planning_ for_change_2008.pdf (accessed 30 May 2013)>.

Villarroel-Walker, R. and Beck, M. (2012). Understanding the Metabolism of Urban-rural Ecosystems, A Multi-sectoral Systems Analysis. *Urban Ecosystems*, 15, 809–848.

Wachsmuth, D. (2012). Three Ecologies: Urban Metabolism and the Society-Nature Opposition. *The Sociological Quarterly*, 53, 506–552.

Zhang, Y., Liu, H. and Chen, B. (2012). Comprehensive Evaluation of the Structural Characteristics of an Urban Metabolic System: Model Development and a Case Study of Beijing. *Ecological Modelling*, 252, 106–113.

2

DECISION SUPPORT TOOLS FOR URBAN PLANNING

Nick Hodges[1], Ainhoa González[2] and Zina Mitraka[3]

[1]NEWCASTLE UNIVERSITY [2]TRINITY COLLEGE [3]FOUNDATION FOR RESEARCH AND TECHNOLOGY - HELLAS

Urban planning and decision support

The origins of urbanism are in Catalhoyuk (Anatolia 7000–5500 BC) and Mesopotamia where urban settlements began to develop between 5500–3500 BC. The initial *ad hoc* development focused on streets. As generations increasingly migrated from rural hinterlands towards cities, pressures built up requiring the introduction of trading to sustain the growing community of citizens. This led to the provision of housing, infrastructure, utilities, services, environment, etc. By 650 BC, Nineveh became the world's largest city with a population of over 120,000. The phenomenon spread from modern day Iraq and the Persian Gulf towards Egypt (e.g. Alexandria had over 300,000 inhabitants by 60 BC) in the West, and the Indus Valley and China in the East. In Mesoamerica, the Mexican City of Teotichucan reached 40,000 by 100 BC. Athens and Rome became models for many early European cities. The rate of migration has steadily increased resulting in approximately 3 per cent of the world's population being city dwellers in the 1800s. Today at 3 billion the figure is more than 50 per cent, and it is expected to reach 75 per cent by 2050. Currently 80 per cent of European citizens live in cities and towns, many of which are characterised by intense road traffic and air pollution.

During the late 1800s roads became congested with horse-drawn vehicles, and public health concerns placed great demands on the infrastructure. The arrival of canals and railways widened the hinterland and led to the development of suburbs leading to further environmental pressures because of daily commuting and smoke emissions. This was exacerbated by the advent of the motor vehicle. With the ready availability of labour and better communications, industry sprang up attracting new migration, increasing resource demand and thus adding to the already major environmental pressures.

Whilst initially traffic congestion could be managed using police officers and then traffic signals, gradually transport planning became more formalised until by the 1960s help was needed to manage the conflicting demands on a day to day basis. In the transportation field, embryonic urban traffic control and management systems began to appear. With the introduction of computers and Information Communications Technologies (ICT) the systems became more sophisticated so that Intelligent Transport Systems (ITS) could offer decision support, in both strategic and real-time horizons. An example of a local authority based transport and environmental integrated system initially developed via the

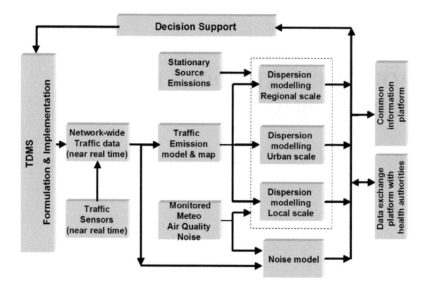

FIGURE 2.1   The HEAVEN Decision Support System

European Union (EU) 6th Framework Programme (FP6) Project "Healthier Environment through the Abatement of Vehicle Emissions and Noise" (HEAVEN, 2007) is illustrated in Figure 2.1.

In the HEAVEN Decision Support System (DSS), archive and real-time data are integrated with traffic, meteorological, air quality and noise modelling to support the highway network manager in selecting Traffic Demand Management Strategies (TDMS). This optimises the movement of traffic and its impact on air and noise quality using the internet to deliver data and information to the operator and the public. This infrastructure has been further enhanced using Co-operative Vehicle Highway Systems (CVHS) so that it forms the transportation core of a smart city (see Figure 2.2).

Development and land use planning imposed new demands on infrastructure, utilities and services provision with their associated impacts on the environment and the social fabric of communities. The new breed of planners from widely differing disciplines were in danger of being overwhelmed by the ever increasing range of competing demands which need to be accommodated within an integrated system. With the move towards widespread public consultation and real community involvement in the decision processes, some sort of methodology was required to help with achieving transparent and equitable decision making. A user friendly DSS which can help both the professional and the layperson to cope with the decision making required to achieve optimum solutions was therefore needed. DSS provide tools for compiling data and knowledge, as well as public/stakeholder perceptions and for processing this information in order to present, compare and rank alternatives and, ultimately, select the one that satisfies the established decision criteria (Carsjens and Ligtenberg, 2007).

Significant progress has been made over the past decade by the green/sustainable building industry in tracking energy and material flows at the building scale. The challenge has been to design sustainable neighbourhoods and cities by directly influencing their urban metabolism processes. Recent advances in bio-physical sciences have led to new methods and models to estimate local scale energy, water, carbon and pollutant fluxes. However, there is often a lack of communication of this new knowledge and its implications in an easily understandable format to end users, such as urban planners, architects and engineers. There has been increasing attention to bridge this gap and towards a better integration of scientists' knowledge into planning.

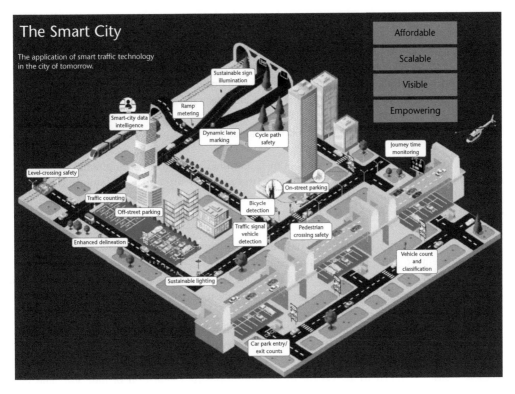

FIGURE 2.2 The Smart City – Transportation aspects

Source: Ladyman, 2013.

Key components of a decision support tool

The report *Sustainable Urban Development in the European Union: A Framework for Action* (EC, 1998) and *The Leipzig Charter* (European Council, 2007) have laid out the principles and strategies towards sustainable urban development policy to be followed by national and local governments. Several national policy and guidance documents have put sustainable development as the core objective of planning. As a result, evaluation procedures are necessary to assess if local urban development initiatives can contribute to progress towards the national goals of sustainable development. Planning evaluation methods have evolved from Environmental Impact Assessment (EIA) to Strategic Environmental Assessment (SEA) and Social Impact Assessment (SIA), with techniques such as Cost Benefit Analysis (CBA) to planning balance sheets and Multi-Criteria Analysis (MCA). This shift has resulted in a move to scientifically and technically more sophisticated methods: from "simple" calculations to complex assessment frameworks; from an environmental focus to an integrated sustainability agenda; and from an aggregated or reductionist strategy to a disaggregated and multidimensional approach (Khakee, 2003). The sustainable urban development evaluation frameworks need to respond to several requirements in order to be effectively applied as decision and design support tools to urban design practice in the *ex ante* evaluation of design proposals. They need to:

• have an integrated conception of sustainable urban development;
• reflect a widely accepted vision that provides guidance during the design process;

- agree objectives and targets to work towards, instead of comparing to the reference baseline scenario;
- allow for early stage deployment, when few data on "project" are available;
- use disaggregate measures and include MCA features;
- offer interaction with the design and be sensitive to design changes;
- allow for (re)iteration, assessing alternatives and supporting the evolution of the design;
- offer communication methods that make the results clear and understandable to the various stake-holders; and
- assess the planning process itself in terms of dialogue and participation of the various stakeholders.

Gil and Duarte (2013) provide a review of over 35 sustainable urban development evaluation tools. The value of these is increasing with availability of large scale systems for data and simulation. The paper suggests that selected evaluation tools are classified according to one the following tool types, adapted from Jensen and Elle's typology (2007):

- *Design guides* – descriptive collections of sustainable urban development themes; checklists as practical instruments to guide the design process.
- *Calculation tools* – software tools for the direct calculation of sustainable urban development indicators; facilitate aggregation of indicators for visualisation in simple charts and/or display thematic maps of individual indicators.
- *Assessment tools* – advanced checklists with software implementation; values for a sustainable urban development theme of structural evaluation framework, and results plotted in charts providing a visual and quantitative profile of different design options.
- *Rating systems* – similar to assessment tools; require precise calculation of indicators and include target values and weights for aggregating the results into the final score.

Gil and Duarte (2013) proposed a general structure for sustainable urban development evaluation tools consisting of the following *five hierarchical levels* with increasing detail and specificity:

- *Sustainability dimensions* – the core goals of sustainability, often based on the three pillars of environment, society and economy. The triple bottom up line.
- *Urban sustainability issues* – the themes of concern to sustainable urban development, which need to be addressed to achieve the core goals (e.g. resources, accessibility, viability).
- *Evaluation criteria* – the aspects that need to be assessed in order to verify the response of the plan to the issue (e.g. energy consumption, waste production, access to public transport or access to jobs).
- *Design indicators* – measurements that are indicative of the performance of the design, with specific measurements and methods (e.g. percentage of residents within 'X' metres walking distance).
- *Benchmark values* – reference or target values that the indicators need to meet to achieve specific quality levels (reference values come from a baseline assessment of similar cases, while target values are objective goals from a more universal sustainability vision).

The main benefits of DSS include: improved data structuring and management, creation of new evidence to support decision making, exploration of personal knowledge and preferences, promotion of learning, and more informed problem solving. The EU 7th Framework Programme (FP7) Project BRIDGE (Chrysoulakis et al. 2013) was created to explore the challenges to be overcome to develop a Geographic Information Systems (GIS) based DSS. ICT tools are also incorporated into the DSS in the form of models, to provide quantitative and generally spatial measures of energy, water, carbon and pollutant fluxes, as well as to enable the incorporation of public values and perceptions. The project set out to initially explore sustainable urban planning by accounting for urban metabolism focusing on the integration of these four key topic areas of energy, water, carbon and pollutants. Later chapters describe the system in greater detail. These illustrate that the BRIDGE DSS:

- responds to the several requirements of sustainable urban development evaluation frameworks;
- deploys calculation and assessment tools and rating systems; and
- operates within Gil and Durarte's five hierarchical levels.

References

Carsjens, G.J., and Ligtenberg, A. (2007). A GIS-based support tool for sustainable spatial planning in metropolitan areas. *Landscape and Urban Planning*, 80: 72–83. DOI: 10.1016/j.landurbplan.2006.06.004

Chrysoulakis, N., Lopes, M., San José, R., Grimmond, C.S.B., Jones, M.B., Magliulo, V., Klostermann, J.E.M., Synnefa, A., Mitraka, Z., Castro, E., González, A., Vogt, R., Vesala, T., Spano, D., Pigeon, G., Freer-Smith, P., Staszewski, T., Hodges, N., Mills, G., and Cartalis, C. (2013). Sustainable urban metabolism as a link between bio-physical sciences and urban planning: the BRIDGE project. *Landscape and Urban Planning*, 112, 100–117.

EC (1998). *Sustainable Urban Development in the European Union: A Framework for Action*. European Commission, Brussels, Belgium (http://ec.europa.eu/environment/urban/pdf/framework_en.pdf)

European Council (2007). *The Leipzig Charter*. European Council, Brussels, Belgium. (http://ec.europa.eu/regional_policy/archive/themes/urban/leipzig_charter.pdf)

Gil, J., and Duarte, J.P. (2013). *Tools for Evaluating the Sustainability of Urban Design: A review*. Institution of Civil Engineers, UK. DOI: 10.1680/udap.11.00048

HEAVEN (2000). Healthier Environment through the Abatement of Vehicle Emissions and Noise project. EU DGINFSO FP6 (http://www.transport-research.info/web/projects/project_details.cfm?id=36410)

Jensen, J.O., and Elle, M. (2007). Exploring the use of tools for urban sustainability in European cities. *Indoor and Built Environment*, 16(3): 235–247. DOI: 10.1177/1420326X07079341

Khakee, A. (2003). The emerging gap between evaluation research and practice. *Evaluation*, 9(3): 340–352. DOI: 10.1177/13563890030093007

Ladyman, S. (2013). Empowering traveller choice in the Smart City. ITS-UK Local Authority Interest Group Conference, 2013, Clearview Traffic Group.

3

THE BRIDGE APPROACH

Nektarios Chrysoulakis

FOUNDATION FOR RESEARCH AND TECHNOLOGY - HELLAS

Introduction

This chapter aims to outline the approach taken in the FP7 (European Union's 7th Framework Programme for Research and Technological Development) project BRIDGE (sustainaBle uRban plannIng Decision support accountinG for urban mEtabolism) to define urban metabolism based on energy, water, carbon and pollutant fluxes and to develop a Decision Support System (DSS) for sustainable urban planning, which takes account of urban metabolism. The BRIDGE DSS was built as an integrated assessment tool that can be used for sustainable urban planning by exploiting the recent advances in bio-physical sciences that have led to the development of new methods and models to estimate the local scale energy, water, carbon and pollutant fluxes. Often there has been a failure in communicating this new knowledge and its implications in an easily understandable format to end-users, such as urban planners, architects and engineers. The BRIDGE project has highlighted the need to bridge this gap, by integrating of scientific knowledge into the planning process. As planners need to consider environmental and socio-economic issues and impacts simultaneously, evaluation methods and tools need to be integrated to address multiple aspects within decision making regarding sustainable urban planning. In the light of such competing demands, tailored decision-making tools are needed to comprehensively analyze baseline information and anticipate potential impacts, as well as to satisfy multiple-scale, multi-period, multiple-objective and multiple-user needs.

Most urban metabolism studies to date use coarse or highly aggregated data (i.e. top-down approach), often at the city or regional level, that provide a snapshot of resource or energy use, that can't be correlated with specific locations, activities, or people. The inputs and outputs of food, water, energy and pollutants have been studied across multiple cities (Kennedy et al. 2007), and at the scale of the individual city (e.g. Ngo and Pataki 2008). An alternative approach is the 'disaggregated approach' (i.e. bottom-up approach), which involves detailed data (or initially disaggregated data) being used (Pincetl et al. 2012); for example scaling up from individual properties to a neighbourhood (e.g. Codobah and Kennedy 2008: Christen et al. 2011). By relating the spatially explicit flows with the relevant census data and human activities, the inputs and the associated outputs generated can be assessed. In BRIDGE, a disaggregated approach was used with the four-dimensional exchange (time and space) and transformation of energy and matter among small areas of the city and its environment.

Significant progress has been made over the past decade by the sustainable building industry in tracking energy and material flows at the building scale. The challenge ahead is to design sustainable neighbourhoods and cities by directly influencing their urban metabolism processes (Kennedy et al. 2011). This is particularly relevant for:

- Energy: optimize energy efficiency of settlements; maximise efficient use of energy through building services and energy supply; maximize share of renewable energy sources; maximize the use of eco-friendly and healthy building materials.
- Water: minimize water consumption; minimise impairment of the natural water cycle; optimize water recycling and reuse.
- Carbon and pollutants: minimise emissions to the atmosphere; maximize carbon stock and pollutant sinks.

Energy enters, passes and leaves the urban system in several ways and in several physical states and forms. Fuels, electricity, radiation, convective and latent heat are the main categories, but construction materials, food, water and waste also contain stored energy. The aspects of energy which are of interest depend on the scope of concern: urban planners, city administrations, economists, statisticians, meteorologists and physicists, each have a different perspective. For example, the interest of city administrations may relate primarily to optimization of energy fluxes for people use, to address pragmatic questions such as how energy consumption can be influenced; whereas meteorologists are concerned with understanding how energy is transported and stored in urban built structures, to influence urban climate. Often not all exchanges are addressed; for example, urban planners may omit radiation as a heating source, neglect anthropogenic heat contribution to emitted radiation, or specify atmospheric heat fluxes as losses from the system. Micrometeorologists include these losses, as anthropogenic heat flux, an input to the urban energy balance (Chrysoulakis et al. 2013).

Sustainable water management techniques are emphasized in a large number of Urban Water Balance (UWB) studies. The transport and removal of water through the piped water system adds an anthropogenic component to the cycle. The UWB, similar to the urban energy balance, applies the principle of conservation to the transfer (or fluxes) of water through a specific area or catchment, allowing understanding of the spatial and temporal patterns of water availability and use. The UWB is directly linked to the surface energy balance as the mass of evapotranspiration is equivalent to the energy term for latent heat flux (Mitchell et al. 2007).

Similarly, the Urban Carbon Budget (UCB) is constrained by the conservation of mass. Like energy and water, the spatial boundaries of an urban area provide the constraints for quantifying the inputs and outputs (e.g. emissions into the atmosphere), as well as storage changes within the system. Vertical fluxes of carbon dioxide (CO_2), measured or modelled in the atmosphere, provide the integrated result between CO_2 uptake by urban vegetation and emissions from combustion and respiration (Christen et al. 2011).

For pollutants, conservation of mass is fundamental, with the concentrations of atmospheric pollutants regulated by the balance between sources and sinks. The emission, dispersion, transformation and removal processes are influenced by a wide range of factors at different temporal and spatial scales. Despite major technical advances in engine technology, exhaust filtering fuel composition and demand management, traffic remains one of the major sources of contamination in urban areas (Borrego et al. 2011; San José et al. 2012).

The main goal of the BRIDGE DSS was to provide a structured assessment of urban metabolism processes (restricted to energy, water, carbon and pollutants) in different planning alternatives and to provide methods for comparative analysis, ranking and selection, in support of planning decisions. In the following sections, an overview of the BRIDGE approach and outcomes is given. The details of all the aspects of the research can be found in BRIDGE website (www.bridge-fp7.eu).

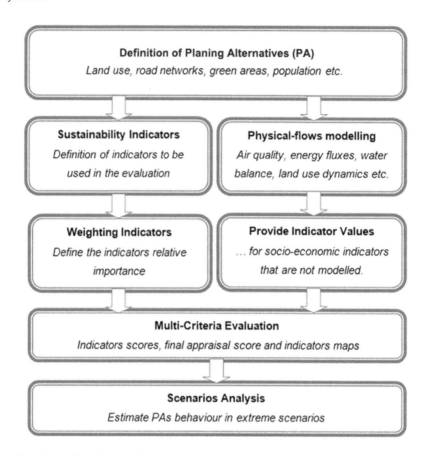

FIGURE 3.1   Flowchart of the BRIDGE methodology

The BRIDGE method

Figure 3.1 shows how all the elements of BRIDGE came together step by step. The following steps are described in more detail: the involvement of users, the different steps that were taken to prioritize specific sustainability issues, to select, implement and evaluate environmental models, to consider planning alternatives and future scenarios and to design and develop the BRIDGE DSS.

The role of end-users in planning alternatives, objectives and indicators definition

Five case study cities were selected, each located in a different part of Europe with different characteristics: a high latitude city that requires a substantial amount of energy for heating (Helsinki, Finland); a low latitude Mediterranean city that requires a substantial amount of energy for cooling (Athens, Greece); a mega-city (London, United Kingdom); a historic city (Firenze, Italy); and an Eastern European city that has undergone significant socio-economic changes in the last decades (Gliwice, Poland).

A 'Community of Practice' (CoP) participatory approach (González et al. 2011) was used to facilitate the interaction between urban planners and BRIDGE scientists (see Chapter 12 for details). CoPs created a learning environment for both groups to search for opportunities to improve sustainable urban planning. A part of each city that needed intervention and 'real life' planning alternatives were identified, as shown in Plate 1:

- In Helsinki, the Meri-Rastila area was chosen. This is a suburban forested peninsula predominantly inhabited by immigrants. The amenity of the area and the need for additional housing and public transport facilities were important topics for this area. The three possible planning alternatives suggested had different type, density and layout of residential buildings.
- In Athens, the study focused on the municipality of Egaleo which has problems with thermal comfort and air quality. The three alternatives considered were: (1) applying cool materials on all buildings and roads in the Egaleo municipality; (2) changing the land use of the Eleonas area with the municipality from its current brownfield/industrial use to urban fabric, including housing and new roads; and (3) changing Eleonas from brownfield to green space.
- In London, the Central Activity Zone (CAZ) was the area of focus. The CAZ contains the central government offices, financial and business services, activities associated with tourism, culture and entertainment, and two important shopping centres. The alternatives were: (1) adding more street trees; (2) adding green roofs; and (3) adding both street trees and green roofs.
- In Firenze, the area of interest consisted of one existing green space (Cascine Park) within the city centre and a former industrial area (San Donato Park). The alternatives entailed the: (1) complete reforestation of a green area and building a sport arena in Cascine Park, including an increase of trees by about 27 per cent of the total; (2) redevelopment of a former industrial area in San Donato Park; and (3) implementation of both alternatives.
- In Gliwice, enhancing the economy by upgrading the urban infrastructure was seen as the most important step. The alternatives considered construction of: (1) a large scale sports facility; (2) a new technologies centre; and (3) both.

CoPs also facilitated the collection of socio-economic data in BRIDGE case studies as described in Chapter 13. This data was combined with environmental observations, as explained in Chapter 14. The BRIDGE approach is based on sustainability objectives and associated indicators addressing specific aspects of urban metabolism. During CoP meetings, planning priorities and core sustainability objectives were determined for each case study. Based on these, indicators were identified by participants and adjusted to the specific requirements of the planning alternatives that were analyzed. These indicators were employed to evaluate how the suggested planning alternatives would modify the fluxes of energy, water, carbon and pollutants in the case studies. Improving air quality was considered to be one of the common objectives, followed by the need to improve energy efficiency and the adaptation to climate change. The objectives, criteria and indicators used in BRIDGE DSS are described in detail in Chapter 15.

Environmental observations

To evaluate the bio-physical models in BRIDGE case studies *in situ* observations were performed. These measurements included standard meteorological data and direct observations of energy, water and CO_2 fluxes using the Eddy Covariance (EC) approach (see details in Chapters 4 and 5). Some sites had already been collecting the data and therefore had a longer time series, e.g. of the order of five years in Firenze (Matese et al. 2009) and Helsinki (Vesala et al. 2008); whereas in London (Kotthaus and Grimmond 2012) and Gliwice (Feigenwinter et al. 2012), the data sets were shorter, as they were established in the framework of BRIDGE. Other environmental variables were collected routinely, or during intensive campaigns to provide additional relevant information for each case study. For example, a two-wavelength Light Detection and Ranging (LiDAR) was deployed in Firenze to provide dust profiles and planetary boundary layer height several times in a day and bio-physical assays of air quality were made using moss bags (Matese et al. 2009). In Helsinki the runoff in two contrasting catchments were instrumented. In Athens, measurements included spatial variations in air temperature, wind and pollutant profiles, indoor environmental quality and impact of the heat island effect on the cooling loads of buildings (Synnefa et al. 2010). In London, surface characteristics were obtained by LiDAR measurements to map tree coverage (Lindberg & Grimmond 2011) and provide the basis for calculating new street tree locations; air quality analysis used the London Air Quality Network (Fuller et al. 2009) and leaf scale observations (Tallis et al. 2011).

Furthermore, remote sensing data were used to derive information about topography, urban landscape and urban form, as well as about the surface and atmospheric geophysical parameters needed in numerical modelling of energy, water, carbon and pollutant fluxes (Chrysoulakis et al. 2013). The Earth Observation activities in the framework of BRIDGE are presented in more detail in Chapter 6.

Simulations of energy, water, carbon and pollutant fluxes

The major modelling effort in BRIDGE presented several challenges (Chrysoulakis et al. 2013): the need to downscale the model results to a scale relevant for urban planning; the need to connect models for different environmental components such as energy, CO_2 and water; the need to respect the constraints in computing time and to ensure the validity of model outputs; and the need to provide comparable reference outputs for the planning alternatives in all case studies in the same time frame. From meso-scale air quality to urban canopy, models were combined using a cascade modelling technique from large to local scale to estimate energy, water, carbon and pollutant fluxes. To determine future distribution of city-wide land uses a cellular automata model (Blecic et al. 2009) was used to account for the broader effects of planning decisions.

A meso-scale meteorological model, linked to two chemical transport models, was used to simulate the meteorological variables and the atmospheric chemistry, based on lumped carbon mechanisms and a detailed description of the photochemistry (see Chapter 7 for details). Aerosol models were used to estimate primary and secondary particulate matter concentrations in the atmosphere. Furthermore, a Regional Climate Model was used for the climate evolution for future scenarios and to provide climate variables and fluxes under climate change. The urban air quality models used to simulate pollutant fluxes are described in more detail in Chapter 8, whereas the urban models used to simulate the energy, water, carbon and pollutant exchanges, for each planning alternative, are presented in Chapters 9, 10 and 11, respectively.

The planning alternatives evaluation method and the DSS

A Multi-Criteria Evaluation (MCE) approach was used to address the complexity of urban metabolism issues reflected in the wide set of sustainability indicators. The MCE enabled comparison and ranking of different urban planning alternatives, through the structured prioritization of a set of nested variables, concerning specific components of urban metabolism: sustainability objectives, criteria and indicators defined for energy, water, carbon and pollutant fluxes. A set of criteria was associated to the objectives. These criteria provide a link between the objectives and the indicators and usually have targets and/or thresholds associated with them. The indicators demonstrate the level of achievement of each criterion in a quantified manner. The indicator's performance is communicated by means of scores determined for each alternative, based on previously defined criteria. MCE enables the weights and scores for all indicators to be combined into a total assessment index. This summary score provides one basis to rank alternatives from best to worst (González et al. 2013). The BRIDGE impact assessment framework is described in more detail in Chapter 15.

MCE was also used for future scenario analysis. Three future scenarios were defined in BRIDGE concerning the exogenous development of the world in terms of sustainability dimensions (Chrysoulakis et al. 2013): I – BRIDGE in wonderland; II – climate change is a burning issue; III – lack of energy is freezing the economy. Different weights were defined by end-users for each of these scenarios. Indicators were remodelled and recalculated for the year 2030. For these projections, assumptions on environmental conditions, based on the Intergovernmental Panel on Climate Change (IPCC) scenarios A2, A1F1 and B1, were used by the BRIDGE models to simulate the fluxes for Scenarios I, II and III, respectively. The strategic scenario analysis that was performed in the framework of BRIDGE is presented in Chapter 17.

The BRIDGE DSS is described in Chapter 16. It consists of the following modules: the Geographic Information System (GIS) module integrates the spatial data (input for the biophysical models, input decision-making procedures, model results) and visualization tools; communication modules as middleware between the GIS and biophysical models; MCE module for the planning alternative assessment; and Graphical User Interface (GUI) that provides the interaction point for the end-users (Chrysoulakis et al. 2010). This GUI leads the end-user through specific steps to produce: indicator maps for each planning alternative; spider diagrams that show the comparative performance of each alternative for each sustainability objective; and a total assessment index for each alternative.

Overview of the BRIDGE outcomes

BRIDGE defined urban metabolism by means of energy, water, carbon and air pollution fluxes at local scale; examined how the change of land use and resources use affects these fluxes; developed indicators to quantify their impacts; developed a DSS based on these indicators; used this DSS to evaluate urban planning alternatives; and supported the development of sustainable planning strategies based on these evaluations. The results of the project are discussed in detail in Parts II, III and IV of this book. A summary of the main BRIDGE outcomes is given in below.

Fluxes measurements and simulations

A common core of measurements had been performed in all case studies. These concern mainly meteorological data and the turbulent exchange of mass and energy as measured by city adapted EC systems. Besides the common batch of meteorological and flux measurements, numerous site specific studies were carried out by means of either environmental parameters collected routinely, or intensive campaigns. Helsinki was the only city where soil water, CO_2 content and flux measurements, as well as storm water quantity and quality, were carried out. In London, an independent measurement of sensible heat flux was performed by means of scintillometry and used for comparisons with EC observations. A detailed presentation of observation results in all case studies is given in Chapter 5.

Concerning fluxes simulations, a system of local climate modelling with high resolution (0.2 km) was successfully implemented in BRIDGE. An integrated scheme was used based on the meso-scale Weather Research and Forecasting (WRF) model (including the urban canopy component), driven by boundary conditions derived by a global general circulation model. Showing great sensitivity to the changes, this scheme allowed urban simulations for analysis and study of the urban metabolism. As discussed in more detail in Chapter 7, the limitations in computer time were the main cause for not having run the high spatial resolution runs following the nesting rate approach of three times as required for numerical and stability reasons. A controversy exists over very high spatial resolution WRF runs (0.2 km, 0.6 km) and numerical issues are raised together with a substantial increase of vertical layers (San José et al. 2012). Several urban canopy models capable of simulating energy, water carbon and pollutant fluxes at local scale were parameterized using WRF outputs. These local scale simulation results are presented in Chapters 8, 9, 10 and 11.

Spider diagrams and final assessment indices

The DSS spider diagrams show the score of each alternative for each objective considered by comparing it to a reference baseline. The spider diagram is represented as a circle with a score of 1 for the baseline of all dimensions, as shown in Plate 2 (Chrysoulakis et al. 2013). Within the spider diagram, indicator scores larger than 1 indicate better performance. The scores for each objective are also integrated into one final assessment index. A final assessment index greater than 1 indicates better performance of the alternative in question compared to the reference. In the example of Helsinki, assuming equal weights for all sustainability dimensions, objectives and indicators (default case), the second planning alternative performed better, even though it had a lower score for the 'water balance' dimension. For Athens, the second alternative performed better, although the first alternative obtained the highest score for the dimension 'thermal comfort'. For London and Firenze the 1st alternative had the highest final assessment index; and for Gliwice, although the first alternatives performed better, the second and third alternatives obtained higher scores for 'thermal comfort' and 'land use' dimensions.

Indicator maps

The impact of different planning strategies (based on changes of urban form) to the energy, water, carbon and pollutant fluxes is illustrated by BRIDGE indicator maps. In these DSS outputs the changes in urban metabolism characteristics caused by each urban planning alternative are indicated as differences from the baseline. The DSS outputs are discussed in more detail in Chapter 16; however, an example of an indicator map for the case study of Athens is given in Plate 3.

In Athens, thermal comfort was addressed. The implementation of each planning alternative would change the surface characteristics, which modifies the energy fluxes (net all-wave radiation flux, turbulent sensible and latent heat fluxes) and therefore air temperature. In Plate 3 (Chrysoulakis et al. 2013), the mean evening (20:00–23:00 LST) air temperature for summertime for the Athens case study (municipality of Egaleo) is shown. During this part of the day, the urban heat island impacts on thermal comfort are typically higher. Planning alternative 1 would have a positive impact relative to the base case air temperature, which is greater over the residential area, of Egaleo. Alternative 1 would reduce the summertime evening air temperature by approximately 0.5K. The modification of the urban energy budget caused by the use of cool materials in the residential area of Egaleo reduced the energy stored in the building materials and, therefore, less energy is transported to the atmosphere as turbulent heat; consequently, the air temperature values were lower than those of the base case. This reduction of air temperature during the evening hours was considered beneficial for the comfort of residents, as well as for the energy consumption for cooling, with obvious socio-economic impacts. The second planning alternative slightly increased the summertime evening air temperature over the brownfield of Eleonas when this was converted to a residential area. However, a small but measurable decrease over the residential area of Egaleo was also observed, which may have been caused by advection. The third planning alternative strongly decreased (around 1.5K) the summertime evening air temperature over the brownfield of Eleonas, which was converted to a green area in this alternative. A small but detectable decrease over the residential area of Egaleo was also observed.

Evaluation of planning alternatives against different future scenarios

The BRIDGE DSS enables the evaluation of urban planning alternatives in the future scenarios context by allowing the end-users to modify the sustainability objectives' and indicators' weights with regards to a specific future scenario and then to generate separate spider diagrams for each scenario. For example, the end-user might prioritize socio-economic benefits over environmental gains in the 'frozen economy' scenario. In the framework of BRIDGE, end-users from the five cities evaluated planning alternatives against BRIDGE scenarios. As is discussed in more detail in Chapter 17, three different types of results were obtained: (1) robust alternatives, which present the best assessment index in all scenarios; (2) unclear evaluation of alternatives, where the indices are very similar, indicating the need to use more detailed information on socio-economic issues; and (3) unstable results according to the future scenarios, which reflects the need to deepen knowledge about future trends, before a decision is taken.

Conclusions and recommendations

The main goal of BRIDGE was to improve the communication of new bio-physical knowledge to end-users (such as urban planners, architects and engineers) with a focus on sustainable urban metabolism. BRIDGE uniquely combined *in situ* measurements of physical flows, high spatial resolution models to simulate these flows, indicators to link the bio-physical processes in urban environment with socio-economic parameters and a DSS to permit evaluation of future development alternatives.

BRIDGE enabled comparisons of planning alternatives' effects on physical flows of urban metabolism aspects. The evaluation of the performance of each alternative was done in a participatory way. This interactive process allowed the end-user to gain an understanding of the relative importance of each sustainability objective and indicator. The combined performance and relative importance of indicators were used to rank planning alternatives in case studies. The DSS was used to assist the end-users to select objectives and indicators and to define their relative importance.

A tool like the BRIDGE DSS may not simplify the urban planning process, but it can help urban planners to deal more adequately with its complexity. Although implementation of the DSS during planning processes may be constrained by a lack of resources and skills at municipalities, practitioners can gain significant insight for more informed decision making. The approach could seamlessly be integrated through a proactive attitude towards sustainability and basic up-skilling of planning staff (e.g. GIS and DSS capacity building), as well as of private sector consultancy in municipalities.

References

Blecic, I., Cecchini, A., & Trunfio, G. A. (2009). A general-purpose geosimulation infrastructure for spatial decision support. *Transaction on Computational Science VI, LNCS*, 5730, 200–218.

Borrego, C., Cascão, P., Lopes, M., Amorim, J. H., Tavares, R., Rodrigues, V., Martins, J., Miranda, A. I., & Chrysoulakis, N. (2011). Impact of urban planning alternatives on air quality: URBAIR model application. In: C. A. Brebbia, J. W. S. Longhurst & V. Popov (Eds): *WIT Transactions on Ecology and the Environment*, Vol 147, WIT Press, ISSN (online) 1743–3541, United Kingdom.

Christen, A., Coops N. C., Crawford B. R., Kellett R., Liss K. N., Olchovski I., Tooke T. R., van der Laan M., & Voogt J. A. (2011). Validation of modeled carbon-dioxide emissions from an urban neighborhood with direct eddy-covariance measurements. *Atmospheric Environment*, 45, 6057–6069.

Chrysoulakis, N., Lopes, M., San José, R., Grimmond, C. S. B., Jones, M. B., Magliulo, V., Klostermann, J. E. M., Synnefa, A., Mitraka, Z., Castro, E., González, A., Vogt, R., Vesala, T., Spano, D., Pigeon, G., Freer-Smith, P., Staszewski, T., Hodges, N., Mills, G., & Cartalis, C. (2013). Sustainable urban metabolism as a link between bio-physical sciences and urban planning: The BRIDGE project. *Landscape and Urban Planning*, 112, 100–117.

Chrysoulakis, N., Mitraka, Z., Diamantakis, E., González, A., Castro, E. A., San José, R., & Blecic, I. (2010). Accounting for urban metabolism in urban planning. The case of BRIDGE. In: CD-ROM of Proceedings of the 10th International Conference on Design & Decision Support Systems in Architecture and Urban Planning, organized by the Technical University of Eindhoven, in Eindhoven, The Netherlands, 19–22 July.

Codobah, N., & Kennedy, C. A. (2008). Metabolism of neighborhoods. *Journal of Urban Planning and Development*, 134, 1–21.

Feigenwinter, C., Vogt, R., & Christen, A. (2012). Eddy covariance measurements over urban areas. In: Eddy Covariance: A Practical Guide to Measurement and Data Analysis. In: M. Aubinet, T. Vesala, & D. Papale (Eds.) *Springer Atmospheric Sciences* (p. 430). Springer, Heidelberg, New York.

Fuller, G., Meston, L., Green, D., Westmoreland, E., & Kelly, F. (2009). *Air Quality in London*. London Air Quality Network Report 14.

González, A., Donnelly, A., Jones, M., Chrysoulakis, N., & Lopes, M. (2013). A decision-support system for sustainable urban metabolism in Europe. *Environmental Impact Assessment Review*, 38, 109–119.

González, A., Donnelly, A., Jones, M., Klostermann, J., Groot, A., & Breil, M. (2011). Community of practice approach to developing urban sustainability indicators. *Journal of Environmental Assessment Policy and Management*, 13, 1–27.

Kennedy, C., Cuddihy, J., & Engel-Yan, J. (2007). The changing metabolism of cities. *Journal of Industrial Ecology*, 22, 43–59.

Kennedy, C., Pincetl, S., & Bunje, P. (2011). The study of urban metabolism and its applications to urban planning and design. *Environmental Pollution*, 159, 1965–1973.

Kotthaus, S., & Grimmond, C. S. B. (2012). Identification of micro-scale anthropogenic CO_2, heat and moisture sources – Processing eddy covariance fluxes for a dense urban environment. *Atmospheric Environment*, 57, 301–316.

Lindberg, F., & Grimmond, C. S. B. (2011). The influence of vegetation and building morphology on shadow patterns and mean radiant temperatures in urban areas: Model development and evaluation. *Theoretical and Applied Climatology*, 105, 311–323.

Matese, A., Gioli, B., Vaccari, F. P., Zaldei, A., & Miglietta, F. (2009). Carbon dioxide emissions of the city center of Firenze, Italy: Measurement, evaluation, and source partitioning. *Journal of Applied Meteorology and Climatology*, 48, 1940–1947.

Mitchell, V. G., Cleugh, H. A., Grimmond, C. S. B., & Xu, J. (2007). Linking urban water balance and energy balance models to analyse urban design options. *Hydrological Processes*, 22, 2891–2900.

Ngo, N. S., & Pataki, D. E. (2008). The energy and mass balance of Los Angeles County. *Urban Ecosystems*, 11, 121–139.

Pincetl, S., Bunje, P., & Holmes, T. (2012). An expanded urban metabolism method: Toward a systems approach for assessing urban energy processes and causes. *Landscape and Urban Planning*, 107, 193–202.

San José, R., Pérez, J. L., Crysoulakis, N., and González, R. M. (2012). WRF-UCM and CMAQ very high resolution simulations (200 m spatial resolution) over London (UK), Athens (Greece), Gliwice (Poland), Helsinki (Finland) and Florence (Italy). In book of abstracts of the 11th Urban Environment Symposium, p. 90, Karlsruhe, Germany, 16–19 September.

Synnefa, A., Stathopoulou, M., Sakka, A., Katsiabani, K., Santamouris, M., Adaktylou, A., Cartalis, C., & Chrysoulakis, N. (2010). Integrating sustainability aspects in urban planning: The case of Athens. In: Proceedings of the 3rd International Conference on Passive and Low Energy Cooling for the Built Environment (PALENC 2010) & 5th European Conference on Energy Performance & Indoor Climate in Buildings (EPIC 2010) & 1st Cool Roofs Conference, 29 September – 1 October 2010, Rhodes, Greece.

Tallis, M., Taylor, G., Sinnett, D., & Freer-Smith, P. (2011). Estimating the removal of atmospheric particulate pollution by the urban tree canopy of London, under current and future environments. *Landscape and Urban Planning*, 103, 129–138.

Vesala, T., Järvi, L., Launiainen, S., Sogachev, A., Rannik, Ü., Mammarella, I., Siivola, E., Keronen, P., Rinne, J., Riikonen, A., & Nikinmaa, E. (2008). Surface-atmosphere interactions over complex urban terrain in Helsinki, Finland. *Tellus B*, 60, 188–199.

PART II

Measurements and modelling of physical flows

4

PHYSICAL FLUXES IN THE URBAN ENVIRONMENT

Björn Lietzke[1], Roland Vogt[1], Duick T. Young[2] and C.S.B. Grimmond[2]

[1]UNIVERSITY OF BASEL AND [2]UNIVERSITY OF READING

Urban metabolism: the meteorological view

Meteorologists are most interested in understanding how energy in the form of radiation and heat influences the urban climate and how this energy is transported, transformed and stored (e.g. in urban building structures). They also are interested in the effects of precipitation on cities, how storm water runoff is changed and how much water is emitted into the atmosphere through evapotranspiration. In addition, they want to know how much cities worldwide contribute to climate change through their emissions to the global carbon cycle. For meteorologists to address the challenges of sustainable cities and urban planning, information on the distribution and flows of energy, water and carbon in typical urban systems have to be known.

From a meteorological perspective, the urban metabolism of a city is strongly dependent on the prevailing regional and local climate and its built-up structure. Together these define the microclimate within the street canyons, on the roads, in the buildings, and at any other place in an urban area. In this context, the urban energy, water and carbon balances are presented in this chapter.

Urban atmosphere

Layers and scales

A key issue of importance for urban investigations is the definition of the appropriate scale of a study area. A classification of urban canopy layer (UCL) elements according to scale considerations is given in Table 4.1. Vertically, the urban atmosphere can be divided into layers as illustrated in Figure 4.1. The lower atmosphere that is influenced by the urban structure is called the urban boundary layer (UBL). From the ground up to roughly the average height of roughness elements like buildings or trees (z_h) is the UCL. It is produced by micro-scale processes in their immediate surroundings. The UCL is part of the roughness sublayer (RSL) which is dependent on the height and density of roughness elements and extends to the height $z \times = a \cdot z_h$ where a ranges between 2 and 5 (Raupach et al. 1991). Above this is the inertial sublayer (ISL) where under ideal conditions vertical fluxes of energy or matter can be expected to be constant with height. The upper part of the UBL, which is to a large extent determined by meso-scale advective processes, is referred to as the outer UBL (Rotach et al. 2005).

TABLE 4.1 Classification of elements of the urban canopy layer (UCL) and their scales (adapted from Oke 2006)

UCL units	Built features	Meteorological scale	Typical horizontal length scales
1. Element	Individual surface element (pavement, trees etc.)	Micro	< 10 m × 10 m
2. Building	Building	Micro	10 m × 10 m
3. Canyon	Street, canyon, property	Micro	30 m × 40 m
4. Block	Block, neighbourhood, factory	Micro/Local	0.5 km × 0.5 km
5. Land-use class (UTZ, UCZ, LCZ, UZE)*	City centre, residential, or industrial zone	Local	5 km × 5 km
6. City	Urban area	Local/Meso	25 km × 25 km
7. Urban region	City plus its environs	Meso	100 km × 100 km

*A number of different classifications at this scale exist including: UTZ: Urban Terrain Zones (Ellefsen 1990/91), UCZ: Urban Climate Zones (Oke 2006), LCZ: Local Climate Zone (Stewart & Oke 2012), UZE: Urban Zones for Energy partitioning (Loridan & Grimmond 2012).

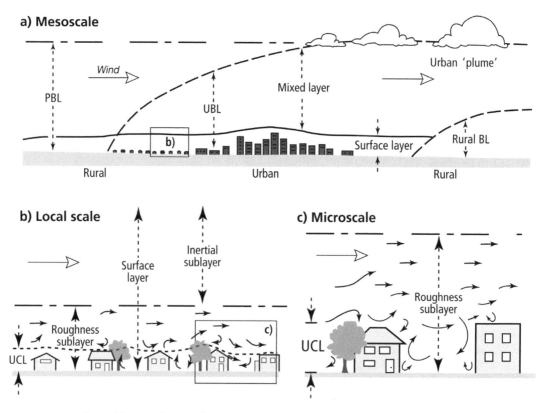

FIGURE 4.1 Scales and layers (planetary boundary layer: PBL; urban boundary layer: UBL; urban canopy layer: UCL) in the urban atmosphere (Feigenwinter et al. 2012; adapted from Oke 2006)

Processes and variability

The exchange of mass and scalars in the urban atmosphere is governed by several processes linked to the heterogeneity of the 3D urban structure. These have a direct influence on the emission and distribution of energy, water and carbon and their transport to the atmosphere. Enhanced mechanical and thermal turbulence in cities change the wind field and induce perturbed streamlines which have an influence on micro- to local-scale transport processes.

Given urban areas are not spatially homogeneous, atmospheric measurements in the UBL are strongly dependent on the spatial and temporal source/sink distribution. This leads to strict requirements for the siting of measurement instruments (Feigenwinter et al. 2012; Oke 2006) as vertical turbulent fluxes, for example, are extremely sensitive to strong local sources in combination with prevailing wind directions (Lietzke & Vogt 2013). Ideal sites are hard to find and it is thus of great importance to know the source area of atmospheric measurements, i.e. the urban area for which observations are representative.

Methods

The energy, water and carbon balances of an urban system can be determined by considering their physical flows in and out of a control volume, which, considering mass conservation, leads to a volume balance approach as depicted in Figure 4.2.

The measurement of the fluxes is achieved with different, often very specific methods. These methods are discussed in the subsequent sections together with the respective processes they measure. One elementary and widely used method to derive the vertical exchange of energy and mass as part of an air volume, the Eddy Covariance (EC) method, is presented here, since this method was mainly used in the BRIDGE project (Chrysoulakis et al. 2013), as described in Chapter 5.

The EC method relies on the fact that atmospheric turbulence is usually the main vertical transport mechanism in the ISL of the UBL. High frequency variations (typically 10–20 Hz) of the vertical wind component w and the scalar s of interest (e.g. H_2O or CO_2) are measured and, after decomposing into mean and turbulent parts by applying Reynolds averaging, their covariance $\overline{w's'}$ gives the vertical turbulent exchange rate of the respective scalar. The primes denote the deviations from the mean and the overbar the average.

Measurements have to be situated at the top of the control volume (Figure 4.2), which is ideally inside the ISL, to capture the vertical transport in and out of the volume. The instrument of choice is usually an ultrasonic anemometer-thermometer in combination with a gas analyzer that measures the scalars of interest (see Chapter 5 for details). An extensive overview on the EC method is given in Aubinet et al. (2012).

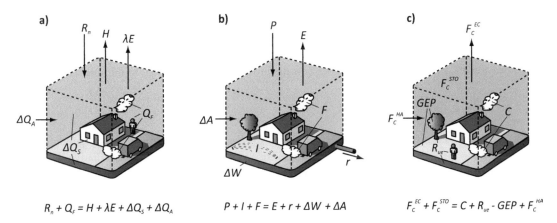

$$R_n + Q_F = H + \lambda E + \Delta Q_S + \Delta Q_A \qquad P + I + F = E + r + \Delta W + \Delta A \qquad F_C^{EC} + F_C^{STO} = C + R_{ue} - GEP + F_C^{HA}$$

FIGURE 4.2 Schematic depiction of the (a) Urban Energy Balance; (b) Urban Water Balance; and (c) Urban Carbon Balance from a micrometeorological perspective. The directions of the arrows represent positive fluxes. For an explaination of variables see the text. (Adapted from Feigenwinter et al. 2012.)

For inhomogeneous urban areas, the EC method is more suitable than other approaches, such as the flux-gradient relations, which normally fail in the RSL (Christen 2005; Piringer et al. 2002; Roth & Oke 1995). Measurements higher up in the ISL are difficult in urban areas due to a lack of higher towers and because of fetch considerations. Therefore, care needs to be taken when using micro-meteorological techniques to consider averaging time, the flux source area and sensor placement to ensure representativeness of the flux in an urban context (Foken 2008; Grimmond 2006).

Physical fluxes

Energy fluxes – Urban Energy Balance

Introduction

Following the volume balance approach, the energy balance of an urban system (Urban Energy Balance – UEB) can be determined by considering the energy flows in and out of the control volume:

$$R_n + Q_F = H + \lambda E + \Delta Q_S + \Delta Q_A$$

where R_n is the net all-wave radiation, Q_F is the anthropogenic heat flux, H is the turbulent sensible heat flux, λE is the turbulent latent heat flux, ΔQ_S is the net storage change within the control volume and ΔQ_A is the net advected flux. All terms are usually expressed as energy flux density per horizontal or vertical area (typically Wm^{-2}, also $MJ\ m^{-2}\ d^{-1}$ for temporal sums). In the following sections each of the UEB terms are discussed.

Theoretical knowledge of the processes forming the UEB and the resultant effects on the UBL is well developed based on numerous observational studies. For typical urban areas, the daytime energy balance is characterized by a significant storage heat flux term, a strong sensible heat flux away from the surface and weak evapotranspiration. As a consequence of strong nocturnal release of stored heat, both turbulent heat fluxes remain directed upward on average at night, a notable difference to the rural environment. This has consequences for the stability of the urban ISL and the RSL which are thermally unstable most of the time (Christen 2005).

Net all-wave radiation R_n

Net all-wave radiation (R_n) is the balance between the incoming ($\downarrow$) and outgoing ($\uparrow$) short- (SW) and long-wave (LW) radiation fluxes and represents the primary source of energy in the UEB:

$$R_n = SW \downarrow - SW \uparrow + LW \downarrow - LW \uparrow$$

Measurements can be made using pyranometers for the short-wave fluxes and pyrgeometers for the long-wave fluxes, or by using net radiometers. In a typical urban atmosphere radiative fluxes are, if compared to their rural counterparts, altered by pollutants. Whereas $SW\downarrow$ will be reduced, $LW\downarrow$ is greater. In typical mid-latitude cities, these changes are normally opposed by a lower short-wave albedo due to darker surface materials (whereas in low-latitude cities, walls and roofs are generally brighter) and a higher surface temperature at night, which augments the long-wave emission (Oke 1987). The net effect on urban/rural radiation differences therefore remains small (Oke 1987; Rotach et al. 2005).

Anthropogenic heat flux Q_F

The anthropogenic heat flux (Q_F) derives mainly from combustion exhausts by stationary and mobile sources (Grimmond 1992; Sailor 2010). Thus, its contribution to the UEB tends to be highest in cold climates in the wintertime when the energy input from human sources is comparatively large (primarily due to domestic heating). But even in summertime it may become significant for cities with high air conditioning usage. Q_F is difficult to determine because of its strongly varying patterns in space and time and because it cannot be measured directly. It is therefore not surprising that many different approaches to estimate this term can be found in literature.

A common approach is to estimate (Q_F) based on inventories of existing socio-economic data, e.g. from energy use data (Sailor 2010). These kinds of data have been analysed as part of the BRIDGE project (Allen et al. 2010; Iamarino et al. 2012; Lindberg et al. 2013; Chapter 5). A second approach, if daily or yearly totals of the energy balance equation are considered, and ΔQ_S can be assumed to be zero, allows calculation of Q_F as the residual term (Christen & Vogt 2004; Pigeon et al. 2007a), or with storage heat flux measurements at a monthly diurnal timescale (Offerle et al. 2005). A third approach explored as part of the BRIDGE project, uses micro-scale analysis of the EC data (Kotthaus & Grimmond 2012) to determine the amount of energy released from buildings. This uses the 'spikes' of heat, water and carbon dioxide (CO_2) 10 Hz data, which impact the departure of the mean used in the EC calculation.

The spatial and temporal patterns of Q_F, have large impacts on the urban climate and is impacted by many of the urban planning alternatives (Chapter 3), therefore understanding the role and size of this term is important.

Turbulent sensible heat flux H

The vertical transport of energy by the sensible heat flux (H) as measured by the EC method is expressed:

$$H = \rho c_p \overline{w'T'}$$

where ρ is the air density (kg m^{-3}), c_p is the specific heat capacity of air (J kg^{-1} K^{-1}) and $\overline{w'T'}$ w (K m s^{-1}) is the average of the product of the turbulent fluctuations of air temperature T and the vertical wind speed. During daytime this term is primarily driven through energy input by R_n, while at night storage release from the urban structure keeps H at a higher level compared to rural areas.

Turbulent latent heat flux λE

The turbulent latent heat flux λE transports moisture away from the surface because of a change of state (e.g. condensation, evaporation). This depends primarily on the availability of water, particularly the presence of vegetated areas (transpiration) or wet surfaces (evaporation). Similar to the sensible heat flux it can be written as:

$$\lambda E = L_v \overline{w'\rho_v'}$$

with L_v the latent heat of vaporization (J kg^{-1}) and ρ_v' the fluctuating water vapor density (kg m^{-3}). λE can be measured directly using the EC method (e.g. a sonic anemometer coupled with an open-path infrared gas analyzer). The quantification of λE is complicated by the extremely heterogeneous sources of moisture. This term is discussed further in the next section, when evapotranspiration (its water equivalent) is considered.

Net storage change ΔQ_S

The rate of change of heat storage (ΔQ_S) consists of the uptake or release of energy by the ground, buildings and vegetation and in the volume. It includes the changes of latent and sensible heat content in the air of the considered control volume. The latter changes are often neglected as they are small compared to the heat storage changes in urban materials.

ΔQ_S within an urban control volume can be theoretically expressed as the sum of storage fluxes for single surface elements (Offerle et al. 2005):

$$\Delta Q_S = \sum_i \frac{\Delta T_i}{\Delta t} (\rho C)_i \Delta x_i \lambda_{pi}$$

where $\Delta T/\Delta t$ is the rate of temperature change, ρC (J m^{-3} K^{-1}) is the volumetric heat capacity, Δx (m) is the element thickness and λ_p (m^2) is the plan area index for each element i (Offerle et al. 2005).

As cities are not expected to cool down, or heat up during a year, the annual total of ΔQ_S has to be zero by definition (Christen 2005; Offerle et al. 2005). This is helpful in calculating annual surface energy balances and in assigning annual residuals to other terms as, for example, the anthropogenic heat flux. ΔQ_S is a spatially and temporally variable term of the energy balance, depending on differences in surface type and radiant loading. It is of particular relevance in the urban energy balance as it can account for more than half of the daytime net radiation at highly urbanized sites (Roberts et al. 2006).

Direct measurements in urban areas are practically unattainable due to the complexity of urban structures and materials. ΔQ_S therefore has to be determined by indirect methods or models. As for most fluxes that are not directly measureable, there is a lack of standard for the determination of urban heat storage and quite a range of methods exist. A commonly used method is to consider the storage flux term as the residual of the energy balance (e.g. Christen & Vogt 2004; Grimmond & Oke 1995, 1999; Roth & Oke 1994; Spronken-Smith et al. 2006):

$$\Delta Q_S = R_n - H - \lambda E$$

ΔQ_A and Q_F are here considered as negligible. Another widely used parameterization approach is based on relations between the net all-wave radiation R_n and the storage heat flux ΔQ_S for typical surface materials (Camuffo & Bernardi 1982; Grimmond & Oke 1991; Oke et al. 1981).

Net advected flux ΔQ_A

Storage change in a control volume due to advection can be expressed as a result of the flow in and out of the volume:

$$\Delta Q_A = Q_A^{in} - Q_A^{out}$$

The scale of the advection is critical relative to the scale of interest. Local-scale advection has largely been neglected for a long time in urban measurement studies based on assuming that the fetch conditions were similar so the term could be considered to be small and the theoretical assumption of horizontal homogeneity was adopted. However, the fetch is rarely sufficiently extensive and consistent, so the latter is often questionable.

To date ΔQ_A has only been investigated at the local scale in urban environments in cities with meso-scale circulations, such as diurnal sea-breeze circulations (e.g. Pigeon et al. 2007b), or drainage flows (e.g. Spronken-Smith et al. 2006) where it has been shown to be important. The circulations between the city and the surroundings (e.g. Lemonsu & Masson 2002) and because of local-scale features (e.g. urban parks; Spronken-Smith et al. 2000) are thought to be important influences in urban areas. However, these processes remain under-studied in urban areas because of the vast array of instrumentation needed and the need to couple the observations with 3D modelling (e.g. Pigeon et al. 2007b). In the BRIDGE project, the role of advection has been considered at the local scale in London (Kotthaus & Grimmond 2013a, 2013b; Loridan et al. 2013).

Water fluxes – Urban Water Balance

Introduction

The urban environment is significantly different to natural hydrological watersheds in terms of land use, water flows and surface cover leading to the modification of the hydrological cycle. In addition, the transport and removal of water through the piped water system adds an anthropogenic component. Artificial surfaces found in urban areas enhance the surface runoff leading to an enhanced risk of flooding and the transport of pollutants (Burian et al. 2002), along with a reduction in infiltration leading to lower replenishment of groundwater (Stephenson 1994).

The Urban Water Balance (UWB) applies the principle of mass conservation to the transfer of water through a specific domain, or catchment (Grimmond et al. 1986), allowing the study of both spatial and temporal patterns of water supply and usage (Mitchell et al. 2001). It can be written as (Grimmond & Oke 1991):

$$P + I + F = E + r + \Delta W + \Delta A$$

where P is precipitation, I is the urban piped water supply, F is water release due to human activity, E is evapotranspiration, r is runoff, ΔW is net change in water storage and ΔA is the net advection of moisture in and out of the control volume. Each of the terms is usually expressed as a depth of water, or as a volume per unit time. It is also common to express individual terms as a percentage of the annual precipitation (often assumed to be the main input into the system) especially in the study of individual components such as runoff and evapotranspiration (e.g. Berthier et al. 2006; Xiao et al. 2007).

Precipitation P

Precipitation is a key input into the UWB as the amount and intensity directly impact the potential magnitude of evapotranspiration, runoff and infiltration and the amount of recharge to surface and groundwater stores. The components of total precipitation (P) are:

$$P = P_r + P_h + P_s + P_m$$

where P_r is rainfall, P_h is hail, P_s is snow and P_m is atmospheric moisture which condenses on contact with the surface in the form of fog, mist or dew. The form of precipitation dictates the timing of the availability of water for runoff, infiltration and evapotranspiration. Snow and hail, which fall in a solid/semi-solid state, have to undergo a change of state to liquid or gaseous form and thus for a time period may be recorded as an increase in storage in the UWB. Depending on the climate, this can last for many months and affect the UWB at a later date through runoff or evaporation (e.g. Järvi et al. 2014, for Helsinki).

Precipitation measurement within urban areas has traditionally used tipping bucket rain gauges. Radar can provide spatial information, but cannot be used alone due to uncertainty in its accuracy (Berne et al. 2004; Vieux & Bedient 2004).

Piped water supply I

The total piped water supply (I) consists of:

$$I = I_U + I_R + I_G + I_S$$

where I_U is the internal residential/commercial/industrial water use, I_R is water used for irrigation, I_G is grey or other reused water and I_S is the leakage to/from the piped network.

The magnitude of the water supplied is driven by a combination of demand from urban inhabitants and supply by the water utility companies or agencies, which is determined by availability of surface and groundwater supplies. Measurement of the supplied water is often from water utility company water meters (e.g. Morris et al. 2007).

Irrigation is a major component of piped water use in urban areas, where seasonal precipitation and weather patterns are particularly variable (Mitchell et al. 2001), with variability in irrigation related to specific weather events (Grimmond & Oke 1986). However, determining the actual amount of irrigation (as with other water usage) is a much more complex problem as it is related to human perception and behaviour (e.g. Arnfield 2003; Grimmond & Oke 1986).

Anthropogenic water release due to combustion F

Anthropogenic water release due to combustion of fuels and from industry consists of:

$$F = F_M + F_I + F_V + F_W$$

where F_M is the release of moisture from air conditioning, heating and cooling applications, F_I is the moisture released from industry, F_V is the moisture released due to combustion of from vehicles and F_W is consumption of bottled water. This term has not been neither widely investigated, nor often considered in UWB models (e.g. Grimmond et al. 1986; Mitchell et al. 2001), but in large cities this term can become more important (Moriwaki & Kanda 2004). In Tokyo, Japan local-scale EC observations over a heavily urbanized area (very little vegetation) displayed significantly large latent heat fluxes (> 100 W m^{-2} and at times greater than observed sensible heat flux) in the summer months, as a result of anthropogenic moisture release from building cooling systems (Moriwaki et al. 2008).

Evapotranspiration E

Evapotranspiration includes evaporation of surface water and transpiration through vegetation of water from the sub-surface vadose zone (Xiao et al. 2007). The term is used interchangeably with evaporation in many studies where it is impractical to separate the two components (Brutsaert 1982):

$$E = E_V + E_T$$

where E_V is evaporation and E_T is transpiration. Its energy equivalent is the latent heat flux Q_E.

Given that water is typically limited at the surface within cities due to high areal fractions of un-vegetated and impervious surfaces, actual evaporation rates are limited by surface controls and energy availability. When water availability is unlimited the theoretical maximum evaporation is typically referred to as potential evaporation which is usually greater than the actual evaporation (Aston 1977). Despite these limiting factors, E can be one of the most important terms in the UWB as a result of complex microclimates, surface storage and irrigation (Berthier et al. 2006; Grimmond & Oke 1986, 1991, 1999; Mitchell et al. 2001).

Urban parks and open water bodies are of particular interest due to the relatively high vegetation cover and greater amount of available moisture resulting in distinct microclimates (the former akin to that of a desert oasis) in comparison to surrounding more built up areas (Hathway & Sharples 2012; Spronken-Smith et al. 2000; Steeneveld et al. 2014). Spronken-Smith et al. (2000) observed that daily total evapotranspiration in a park in Sacramento, USA, was greater than 300 per cent of the total from the surrounding irrigated suburban area.

Observation of evapotranspiration has been undertaken using mini-lysimeters at the micro scale (Oke 1979), while at the local scale, micrometeorological techniques are often applied (e.g. EC). Alternatively when direct measurement is unavailable it can be determined as a residual term of the UWB equation or using the Bowen ratio energy balance (Nouri et al. 2012). Goldbach & Kuttler (2013) found in Oberhausen, Germany, using EC, that absolute daily maximum evapotranspiration varied by up to 90 per cent between urban and suburban areas where vegetated surface fractions were 0.18 and 0.58, respectively. Data sets from 19 EC sites located in urban and suburban areas of 15 cities worldwide indicated a positive relation between the active vegetated index (indices based on vegetated fraction and seasonal leaf-area index (Loridan et al. 2011) and mean midday evapotranspiration, with a stronger linear dependence on observed E rates prevalent when active vegetated index was < 0.43 (Loridan & Grimmond 2012).

Runoff r

Runoff is the flow over the surface and through drainage pipes. It represents water that has not been captured by some intermediate store (e.g. tree canopy, roof or surface storage) or has not infiltrated into sub-surface stores within a particular time period. A greater fraction of impervious surfaces in cities in comparison to rural areas leads to more rapid surface flows often enhanced by drainage networks (Semadeni-Davies & Bengtsson 1999). The increase in runoff can lead to a higher probability of flooding and the transport of pollutants (Burian et al. 2002; Xiao et al. 2007). Urban runoff consists of:

$$r = r_S + r_W + r_O + r_L + r_F$$

where r_S is storm water runoff (through storm drains), r_W is waste water flow (sewer system), r_O is runoff released by snow melt, r_L is surface runoff (e.g. overland flow and roof runoff) and r_F is surface infiltration. The rate and magnitude of runoff are regulated by the rate of precipitation, soil moisture content (influences infiltration), land surface properties (e.g. fraction of vegetation cover and perme-ability), local topography and the design of the drainage system infrastructure.

Runoff is often either modelled in the UWB due to a lack of measured data or the size of the study catchment (Branger et al. 2013; Wang et al. 2008), parameterized using infiltration/runoff coefficients (Hollis & Ovenden 1988) or as a residual (Jia et al. 2001). However, runoff measurement is possible directly using flow meters to determine discharge through a drainage system (Ragab et al. 2003b) or controlled study area (Stephenson 1994; Xiao et al. 2007), water capture to collect and measure roof runoff (Hollis & Ovenden 1988; Ragab et al. 2003a), and indirectly using water balance techniques to determine available water for potential runoff (Inkiläinen et al. 2013). In the BRIDGE case studies, the runoff in two small catchments was observed in Helsinki (see Chapter 10 for details).

Net storage change ΔW

The net change in storage term (ΔW) refers to the change in water storage within the study catchment. Its magnitude is determined by

$$\Delta W = \Delta W_g + \Delta W_m + \Delta W_W + \Delta W_a + \Delta W_n$$

where ΔW_g is the net change in ground water storage, ΔW_m is the net change in soil moisture storage, ΔW_W is surface water storage (e.g. ponds and lakes), ΔW_a is anthropogenic storage (e.g. storm water holding and water butts) and ΔW_n is the net change in snowpack storage.

For large catchments, groundwater within the soil and deeper aquifer(s) can be significant. Techniques to measure soil moisture include tensionmeters (Berthier et al. 2004), gravimetric sampling (Grimmond & Oke 1986) and time domain reflectometry and groundwater levels can be observed through boreholes (Stephenson 1994).

Net moisture advection ΔA

The net moisture advection is the horizontal transport of moisture by atmospheric flow. It is driven by flows at a number of atmospheric scales ranging from micro- and local-scale turbulence to meso-scale circulations (e.g. sea breezes and valley flow). In many UWB studies the net moisture advection is not considered (e.g. Grimmond et al. 1986; Lemonsu et al. 2007).

Carbon fluxes – Urban Carbon Balance

Introduction

Compared to energy and water, the urban balance of carbon – in the form of CO_2 – shows greater deviations from its rural counterpart. Anthropogenic CO_2 emissions, derived from the burning of fossil fuels, are the major net source for global atmospheric carbon (Denman et al. 2007) and cities contribute a great share. Thus, knowledge of the spatiotemporal distribution of sources and sinks in urban environments and the processes that determine atmospheric transport in the UBL is of great importance.

Using a volume budget approach that focuses on surface–atmosphere processes, the Urban Carbon Balance (UCB) can be written as:

$$F_C^{EC} + F_C^{STO} = C + R_{ue} - GEP + F_C^{HA}$$

where F_C^{EC} is the integrative turbulent mass flux density of CO_2, F_C^{STO} is the storage change between the surface and the measurement level, C represents emissions through anthropogenic combustion processes, R_{ue} is the respiration of the urban ecosystem (including from humans), GEP stands for the sink effects due to photosynthesis and F_C^{HA} is the horizontal advection contribution. Terms are usually expressed as CO_2 flux density per horizontal or vertical area (typically µmol m^{-2} s^{-1} or kg m^{-2} a^{-1}).

Turbulent CO_2 flux F_C^{EC}

A common way to determine the turbulent vertical mass flux density of CO_2 (F_C^{EC}) is by the use of the EC method, combining sonic with infrared gas analyzer measurements (a list of urban studies can be found in Lietzke et al. (2014). Two types of gas analyzers are widely used: open path analyzers where CO_2 concentrations are measured instantaneously in the probed air volume (e.g. Moriwaki & Kanda 2004; Vogt et al. 2006) and closed path analyzers where air is sucked through a tube into an enclosed measurement system (e.g. Grimmond et al. 2002; Järvi et al. 2012). The first has the advantage of measuring *in situ* but is sensitive to disturbances of the measurement path, e.g. through rain, dew or dust. The latter measurements are subject to a time lag and an attenuation of the signal, dependent on the length of the tube, but are not influenced by meteorological disturbances (Grimmond et al. 2002; Järvi et al. 2009).

Summing F_C^{EC} over a defined timescale yields the net urban ecosystem exchange (*NuEE*) rate analogous to the net ecosystem exchange (*NEE*) rates of rural ecosystems. The main contrast to non–urban ecosystems is that the urban surfaces generally act as a CO_2 source; consequently F_C^{EC} is nearly always positive. This results in positive *NuEE* values which are usually higher the more urbanized an area is.

Net storage change in the air F_C^{STO}

Fluxes in the RSL are not constant with height (Rotach 2001) and thus a vertical flux divergence over time has to be assumed in the air volume between the urban surface and the measurement level. This is considered in the term F_C^{STO}, which can be determined using representative measurements of the concentration change within the air volume over time (Feigenwinter et al. 2012). In an urban environment, this would need several vertical profile measurements to account for the spatial variability within the EC source area – which is rarely feasible. Similar to ΔQ_S, F_C^{STO} can assumed to be zero over a longer time period. On a diurnal scale it becomes relevant as, for example, nocturnally accumulated CO_2 in the shallow UBL and the street canyons is flushed in the morning, when thermal mixing starts, leading to an overestimation of F_C^{EC} compared to the actual emissions (Feigenwinter et al. 2012).

Combustion C

Anthropogenic emissions through combustion of fossil fuels are the main contributors to the UCB, consisting of:

$$C = C_B + C_V$$

The combustion from buildings (C_B) and vehicular traffic (C_V) can be distinguished by the type of fuel they burn (natural gas, oil or wood for heating versus gasoline or diesel for driving) and the spatiotemporal emission patterns. Source distribution is, as for Q_F, very heterogeneous. While C_V can be considered as a line source on the bottom of the control volume that is primarily dependent on the diurnal/weekly traffic use behavior, C_B generated by heating depends on climate related human activity (heating in winter, air conditioning in summer), has a distinct seasonal cycle (Lietzke et al. 2014) and consists of point sources at certain heights (e.g. chimneys) (Kotthaus & Grimmond 2012). Industry emissions as a part of C_B follow their own patterns that need to be taken into account as appropriate.

Through isotopic analyses of air samples (Clark-Thorne & Yapp 2003; Pataki et al. 2003), the fraction of atmospheric CO_2 generated by either C_B or C_V can be derived. Inventory based approaches using fossil fuel consumption data and traffic density analyses (e.g. Helfter et al. 2011; Ward et al. 2013) can give an estimate of C_B and C_V, or are used as input to model their contributions. Spatiotemporal adequately resolved data is rarely available so that e.g. fuel consumption often has to be scaled down from city to neighbourhood or building scale (e.g. Christen et al. 2011). An indicator of fuel burned for heating purposes can be heating degree days based on outside air temperature and the desired inside air temperature (Lietzke et al. 2014).

Urban ecosystem respiration R_{ue}

Urban ecosystem respiration (R_{ue}) can be separated into respiration of soils and vegetation (R_{SV}), waste decomposition (R_W) and human respiration (R_M):

$$R_{ue} = R_{SV} + R_W + R_M$$

Compared to natural ecosystems, urban R_{SV} is influenced by irrigation and fertilization. R_M depends on the density of people that live or work in an area and, on the basis of an individual, the physiological level of activity (active, resting, sleeping etc.). Moriwaki & Kanda (2004) estimated human body respiration emissions at rest to be 8.87 mg CO_2 s^{-1}.

Gross ecosystem productivity GEP

Gross ecosystem productivity (*GEP*) is a measure of the uptake of CO_2 through photosynthesis from the air. In cities, both *GEP* and R_{SV} are primarily dependent on the surface fraction of vegetation (parks, lawns and trees), its density and type and the local climate which determines the seasonal photosynthesis rate. Productivity of urban vegetation is usually high due to irrigation, higher temperatures, less frost damage (urban heat island) and fertilization e.g. nitrogen oxides (NO_X) deposition (Trusilova & Churkina 2008), but physiological stress due to air pollution may lead to reduced *GEP*. Chamber measurements (Christen et al. 2011) help in estimating soil and lawn activity. In urban areas, photosynthesis is typically not able to compensate for the high CO_2 emissions by combustion (Kotthaus & Grimmond 2012; Lietzke & Vogt 2013), but may have a limiting effect on measured fluxes (Coutts et al. 2007; Kordowski & Kuttler 2010; Ward et al. 2013). Depending on the extent of urbanization, particularly vegetation effects, temporary sink effects can be observed (e.g. Crawford et al. 2011; Ramamurthy & Pardyjak 2011).

Net advection F_C^{HA}

Similar to advection in the UEB and UWB, net horizontal advection of CO_2 (F_C^{HA}) in urban areas is rarely addressed in studies. Results from a number of field experiments in forests (Aubinet et al. 2010) show that there is a large uncertainty in quantifying horizontal and vertical advection fluxes. Both terms are large, are coupled and seem not to cancel each other. To date, it is not known how relevant this is for the urban environment.

References

Allen, L., Lindberg, F. & Grimmond, C. (2010). Global to city scale model for anthropogenic heat flux. *International Journal of Climatology*, 31, 1990–2005.

Arnfield, A. (2003). Two decades of urban climate research: a review of turbulence, exchanges of energy and water, and the urban heat island. *International Journal of Climatology*, 23, 1–26.

Aston, A. (1977). Water-resources and consumption in Hong Kong. *Urban Ecology*, 2, 327–351.

Aubinet, M., Feigenwinter, C., Heinesch, B., Bernhofer, C., Canepa, E., Lindroth, A., Montagnani, L., Rebmann, C., Sedlak, P. & Van Gorsel, E. (2010). Direct advection measurements do not help to solve the night-time CO_2 closure problem: evidence from three different forests. *Agricultural and Forest Meteorology*, 150, 655–664.

Aubinet, M., Vesala, T. & Papale, D., eds (2012). *Eddy Covariance – a Practical Guide to Measurement and Data Analysis.* Dordrecht, The Netherlands, Springer Atmospheric Sciences.

Berne, A., Delrieu, G., Creutin, J. & Obled, C. (2004). Temporal and spatial resolution of rainfall measurements required for urban hydrology, *Journal of Hydrology*, 299, 166–179.

Berthier, E., Andrieu, H. & Creutin, J. (2004). The role of soil in the generation of urban runoff: development and evaluation of a 2D model. *Journal of Hydrology*, 299, 252–266.

Berthier, E., Dupont, S., Mestayer, P. & Andrieu, H. (2006). Comparison of two evapotranspiration schemes on a sub-urban site, *Journal of Hydrology*, 328, 635–646.

Branger, F., Kermadi, S., Jacqueminet, C., Michel, K., Labbas, M., Krause, P., Kralisch, S. & Braud, I. (2013). Assessment of the influence of land use data on the water balance components of a peri-urban catchment using a distributed modelling approach. *Journal of Hydrology,* 505, 312–325.

Brutsaert, W. (1982). *Evaporation into the Atmosphere: Theory, History and Applications.* Dordrecht, The Netherlands, Springer.

Burian, S. J., McPherson, T., Brown, M., Streit, G. & Turin, H. (2002). Modeling the effects of air quality policy changes on water quality in urban areas. *Environmental Modeling & Assessment*, 7, 179–190.

Camuffo, D. & Bernardi, A. (1982). An observational study of heat fluxes and their relationship with net radiation. *Boundary-Layer Meteorology*, 23, 359–368.

Christen, A. (2005). Atmospheric turbulence and surface energy exchange in urban environments, PhD thesis, University of Basel, Stratus. ISBN 3–85977–266.X.

Christen, A., Coops, N., Crawford, B., Kellett, R., Liss, K., Olchovski, I., Tooke, T., van der Laan, M. & Voogt, J. (2011). Validation of modeled carbon-dioxide emissions from an urban neighborhood with direct eddy-covariance measurements. *Atmospheric Environment*, 45, 6057–6069.

Christen, A. & Vogt, R. (2004). Energy and radiation balance of a central European city. *International Journal of Climatology*, 24, 1395–1421.

Chrysoulakis, N., Lopes, M., San José, R., Grimmond, C. S. B., Jones, M. B., Magliulo, V., Klostermann, J. E. M., Synnefa, A., Mitraka, Z., Castro, E., González, A., Vogt, R., Vesala, T., Spano, D., Pigeon, G., Freer-Smith, P., Staszewski, T., Hodges, N., Mills, G. & Cartalis, C. (2013). Sustainable urban metabolism as a link between bio-physical sciences and urban planning: the BRIDGE project. *Landscape and Urban Planning*, 112, 100–117.

Clark-Thorne, S. & Yapp, C. (2003). Stable carbon isotope constraints on mixing and mass balance of CO_2 in an urban atmosphere: Dallas metropolitan area, Texas, USA. *Applied Geochemistry*, 18, 75–95.

Coutts, A., Beringer, J. & Tapper, N. (2007). Characteristics influencing the variability of urban CO_2 fluxes in Melbourne, Australia. *Atmospheric Environment*, 41, 51–62.

Crawford, B., Grimmond, C. & Christen, A. (2011). Five years of carbon dioxide fluxes measurements in a highly vegetated suburban area. *Atmospheric Environment*, 45, 896–905.

Denman, K., Brasseur, G., Chidthaisong, A., Ciais, P., Cox, P., Dickinson, R., Hauglustaine, D., Heinze, C., Holland, E., Jacob, D., Lohmann, U., Ramachandran, S., da Silva Dias, P., Wofsy, S. & Zhang, X. (2007). Couplings between changes in the climate system and biogeochemistry. In: *Climate Change 2007: The Physical Science Basis. Contribution of Working Group I to the Fourth Assessment Report of the Intergovernmental Panel on Climate Change*, Cambridge University Press, Cambridge, UK and New York, USA.

Ellefsen, R. (1990/91). Mapping and measuring buildings in the urban canopy boundary layer in ten US cities. *Energy and Buildings*, 15–16, 1025–1049.

Feigenwinter, C., Vogt, R. & Christen, A. (2012). Eddy covariance measurements over urban areas. In M. Aubinet, T. Vesala & D. Papale, eds, *Eddy Covariance – A Practical Guide to Measurement and Data Analysis*. Dordrecht, The Netherlands,, Springer Atmospheric Sciences, p. 430.

Foken, T., ed. (2008). *Micrometeorology*. Berlin, Germany, Springer-Verlag Berlin Heidelberg.

Goldbach, A. & Kuttler, W. (2013). Quantification of turbulent heat fluxes for adaptation strategies within urban planning. *International Journal of Climatology*, 33, 143–159.

Grimmond, C. (1992). The suburban energy balance: methodological considerations and results for a mid-latitude west coast city under winter and spring conditions. *International Journal of Climatology,* 12, 481–497.

Grimmond, C. (2006). Progress in measuring and observing the urban atmosphere. *Theoretical and Applied Climatology*, 84, 3–22.

Grimmond, C., King, T., Cropley, F., Nowak, D. & Souch, C. (2002). Local-scale fluxes of carbon dioxide in urban environments: methodological challenges and results from Chicago. *Environmental Pollution*, 116, 243–254.

Grimmond, C. & Oke, T. (1986). Urban water-balance 2: Results from a suburb of Vancouver, British-Columbia. *Water Resources Research*, 22, 1404–1412.

Grimmond, C. & Oke, T. (1991). An evapotranspiration-interception model for urban areas, *Water Resources Research*, 27, 1739–1755.

Grimmond, C. & Oke, T. (1995). Comparison of heat fluxes from summertime observations in the suburbs of four North American cities. *Journal of Applied Meteorology*, 34, 873–889.

Grimmond, C. & Oke, T. (1999). Aerodynamic properties of urban areas derived from analysis of surface form. *Journal of Applied Meteorology*, 38, 1262–1292.

Grimmond, C., Oke, T. & Steyn, D. (1986). Urban water-balance 1. A model for daily totals. *Water Resources Research*, 22, 1397–1403.

Hathway, E. & Sharples, S. (2012). The interaction of rivers and urban form in mitigating the urban heat island effect: a UK case study. *Building and Environment*, 58, 14–22.

Helfter, C., Famulari, D., Philipps, G., Barlow, J., Wood, C., Grimmond, C. & Nemitz, E. (2011). Controls of carbon dioxide concentrations and fluxes above central London. *Atmospheric Chemistry and Physics*, 11, 1913–1928.

Hollis, G. E. & Ovenden, J. (1988). The quantity of stormwater runoff from 10 stretches of road, a car park and 8 roofs in Hertfordshire, England during 1983. *Hydrological Processes*, 2, 227–243.

Iamarino, M., Beevers, S. & Grimmond, C. (2012). High-resolution (space, time) anthropogenic heat emissions: London 1970–2025. *International Journal of Climatology*, 32, 1754–1767.

Inkiläinen, E., McHale, M., Blank, G., James, A. & Nikinmaa, E. (2013). The role of the residential forest in regulating throughfall: a case study in Raleigh, North Carolina, USA. *Landscape and Urban Planning*, 119, 91–103.

Järvi, L., Grimmond, C., Taka, M., Setälä, H., Nordbo, A. & Strachan, I. (2014, in review). Development of the Surface Urban Energy and Water balance Scheme (SUEWS) for cold climate cities. *Geoscientific Model Development*, MS No.: gmd-2013-163

Järvi, L., Mammarella, I., Eugster, W., Ibrom, A., Siivola, E., Dellwik, E., Keronen, P., Burba, G. & Vesala, T. (2009). Comparison of net CO_2 fluxes measured with open- and closed-path infrared gas analyzers in an urban complex environment. *Boreal Environment Research*, 14, 499–514.

Järvi, L., Nordbo, A., Junninen, H., Riikonen, A., Moilanen, J., Nikinmaa, E. & Vesala, T. (2012). Seasonal and annual variation of carbon dioxide surface fluxes in Helsinki, Finland, in 2006 – 2010. *Atmospheric Chemistry and Physics Discussions*, 12, 8355–8396.

Jia, Y., Ni, G., Kawahara, Y. & Suetsugi, T. (2001). Development of WEP model and its application to an urban watershed. *Hydrological Processes*, 15, 2175–2194.

Kordowski, K. & Kuttler, W. (2010). Carbon dioxide fluxes over an urban park area. *Atmospheric Environment*, 44, 2722–2730.

Kotthaus, S. & Grimmond, C. (2012). Identification of micro-scale anthropogenic CO_2, heat and moisture sources – processing eddy covariance fluxes for a dense urban environment. *Atmospheric Environment*, 57, 301–316.

Kotthaus, S. & Grimmond, C. S. B. (2013a). Energy exchange in a dense urban environment – Part I: Temporal variability of long-term observations in central London. *Urban Climate* In Press, Corrected Proof, doi:10.1016/j. uclim.2013.10.002.

Kotthaus, S. & Grimmond, C. S. B. (2013b). Energy exchange in a dense urban environment – Part II: Impact of spatial heterogeneity of the surface. *Urban Climate* In Press, Corrected Proof, doi: 10.1016/j.uclim.2013.10.001.

Lemonsu, A. & Masson, V. (2002). Simulation of a summer urban breeze over Paris. *Boundary Layer Meteorology*, 104, 463–490.

Lemonsu, A., Masson, V. & Berthier, E. (2007). Improvement of the hydrological component of an urban soil-vegetation-atmosphere-transfer model. *Hydrological Processes*, 21, 2100–2111.

Lietzke, B. & Vogt, R. (2013). Variability of CO_2 concentrations and fluxes in and above an urban street canyon. *Atmospheric Environment*, 74, 60–72.

Lietzke, B., Vogt, R., Feigenwinter, C. & Parlow, E. (2014, in review). On the variability of carbon-dioxide flux and its controlling factors in a heterogeneous urban environment. *International Journal of Climatology*.

Lindberg, F., Grimmond, C., Yogeswaran, N., Kotthaus, S. & Allen, L. (2013). Impact of city changes and weather on anthropogenic heat flux in Europe 1995–2015. *Urban Climate*, 4, 1–15.

Loridan, T. & Grimmond, C. S. B. (2012). Characterization of energy flux partitioning in urban environments: links with surface seasonal properties. *Journal of Applied Meteorology and Climatology*, 51, 219–241.

Loridan, T., Grimmond, C. S. B., Offerle, B. D., Young, D. T., Smith, T. E. L., Jarvi, L. & Lindberg, F. (2011). Local-scale urban meteorological parameterization scheme (lumps): longwave radiation parameterization and seasonality related developments. *Journal of Applied Meteorology and Climatology*, 50, 185–202.

Loridan, T., Lindberg, F., Jorba, O., Kotthaus, S., Grossman-Clarke, S. & Grimmond, C. S. B. (2013). High resolution simulation of the variability of surface energy balance fluxes across central London with urban zones for energy partitioning. *Boundary-Layer Meteorology*, 147, 493–523.

Mitchell, V., Mein, R. & McMahon, T. (2001). Modelling the urban water cycle. *Environmental Modelling & Software*, 16, 615–629.

Moriwaki, R. & Kanda, M. (2004). Seasonal and diurnal fluxes of radiation, heat, water vapor, and carbon dioxide over a suburban area. *Journal of Applied Meteorology*, 43, 1700–1710.

Moriwaki, R., Kanda, M., Senoo, H., Hagishima, A. & Kinouchi, T. (2008). Anthropogenic water vapor emissions in Tokyo. *Water Resource Research*, 44, *W11424*

Morris, B., Rueedi, J., Cronin, A., Diaper, C. & DeSilva, D. (2007). Using linked process models to improve urban groundwater management: an example from Doncaster England. *Water and Environment Journal*, 21, 229–240.

Nouri, H., Beecham, S., Kazemi, F. & Hassanli, A. M. (2012). A review of ET measurement techniques for estimating the water requirements of urban landscape vegetation. *Urban Water Journal*, 10, 247–259.

Offerle, B., Grimmond, C. S. B. & Fortuniak, K. (2005). Heat storage and anthropogenic heat flux in relation to the energy balance of a Central European city centre. *International Journal of Climatology*, 25, 1405–1419.

Oke, T. (1979). Advectively-assisted evapotranspiration from irrigated urban vegetation. *Boundary-Layer Meteorology*, 17, 167–173.

Oke, T. (2006). Towards better scientific communication in urban climate. *Theoretical and Applied Climatology*, 84, 179–190.

Oke, T., Kalanda, B. & Steyn, D. (1981). Parameterization of heat storage in urban areas. *Urban Ecology*, 5, 45–54.

Oke, T. R. (1987). *Boundary Layer Climates*. Routledge, London.

Pataki, D., Bowling, D. & Ehleringer, J. (2003). Seasonal cycle of carbon dioxide and its isotopic composition in an urban atmosphere: anthropogenic and biogenic effects. *Journal of Geophysical Research*, 108(D23), 4735–4742.

Pigeon, G., Legain, D., Durand, P. & Masson, V. (2007a). Anthropogenic heat release in an old European agglomeration (Toulouse, France). *International Journal of Climatology*, 27, 1969–1981.

Pigeon, G., Lemonsu, A., Grimmond, C., Durand, P., Thouron, O. & Masson, V. (2007b). Divergence of turbulent fluxes in the surface layer: case of a coastal city. *Boundary-Layer Meteorology*, 124, 269–290.

Piringer, M., Grimmond, C., Joffre, S., Mestayer, P., Middleton, D., Rotach, M., Baklanov, A., De Ridder, K., Ferreira, J., Guilloteau, E., Karppinen, A., Martilli, A., Masson, V. & Tombrou, M. (2002). Investigating the surface energy balance in urban areas – recent advances and future needs. *Water, Air and Soil Pollution: Focus*, 2, 1–16.

Ragab, R., Bromley, J., Rosier, P., Cooper, J. & Gash, J. (2003a). Experimental study of water fluxes in a residential area: 1. rainfall, roof runoff and evaporation: the effect of slope and aspect. *Hydrological Processes*, 17, 2409–2422.

Ragab, R., Rosier, P., Dixon, A., Bromley, J. & Cooper, J. (2003b). Experimental study of water fluxes in a residential area: 2. road infiltration, runoff and evaporation. *Hydrological Processes*, 17, 2423–2437.

Ramamurthy, P. & Pardyjak, E. (2011). Toward understanding the behavior of carbon dioxide and surface energy fluxes in the urbanized semi-arid Salt Lake Valley, Utah, USA. *Atmospheric Environment*, 45, 73–84.

Raupach, M., Antonia, R. & Rajagopalan, S. (1991). Roughwall turbulent boundary layers. *Applied Mechanics Reviews*, 44, 1–25.

Roberts, S., Oke, T., Grimmond, C. & Voogt, J. (2006). Comparison of four methods to estimate urban heat storage. *Journal of Applied Meteorology*, 45, 1766–1781.

Rotach, M. (2001). Simulation of urban-scale dispersion using a Lagrangian stochastic dispersion model. *Boundary-Layer Meteorology*, 99, 379–410.

Rotach, M., Vogt, R., Bernhofer, C., Batchvarova, E., Christen, A., Clappier, A., Feddersen, B., Gryning, S.-E., Martucci, G., Mayer, H., Mitev, V., Oke, T., Parlow, E., Richner, H., Roth, M., Roulet, Y., Ruffieux, D., Salmond, J., Schatzmann, M. & Voogt, J. (2005). BUBBLE – an urban boundary layer meteorology project. *Theoretical and Applied Climatology*, 81, 149–156.

Roth, M. & Oke, T. (1994). Comparison of modelled and measured heat storage in suburban terrain. *Beitraege zur Physik der Atmosphaere*, 67, 149–156.

Roth, M. & Oke, T. (1995). Relative efficiencies of turbulent transfer of heat, mass and momentum over a patchy urban surface. *Journal of Atmospheric Sciences*, 52, 1864–1874.

Sailor, D. (2010). A review of methods for estimating anthropogenic heat and moisture emissions in the urban environment. *International Journal of Climatology*, 31, 189–199.

Semadeni-Davies, A. F. & Bengtsson, L. (1999). The water balance of a sub-Arctic town. *Hydrological Processes*, 13, 1871–1885.

Spronken-Smith, R. A., Oke, T. & Lowry, W. (2000). Advection and the surface energy balance across an irrigated urban park. *International Journal of Climatology*, 20, 1033–1047.

Spronken-Smith, R. A., Kossmann, M. & Zawar-Reza, P. (2006). Where does all the energy go? Surface energy partitioning in suburban Christchurch under stable wintertime conditions. *Theoretical and Applied Climatology*, 84, 137–149.

Steeneveld, G. J., Koopmans, S., Heusinkveld, B. G. & Theeuwes, N. E. (2014). Refreshing the role of open water surfaces on mitigating the maximum urban heat island effect. *Landscape and Urban Planning*, 121, 92–96.

Stephenson, D. (1994). Comparison of the water-balance for an undeveloped and a suburban catchment. *Hydrological Sciences Journal*, 39, 295–307.

Stewart, I. D. & Oke, T. R. (2012). Local climate zones for urban temperature studies. *Bulletin of the American Meteorological Society*, 93, 1879–1900.

Trusilova, K. & Churkina, G. (2008). The response of the terrestrial biosphere to urbanization: land cover conversion, climate, and urban pollution. *Biogeosciences*, 5, 1505–1515.

Vieux, B. & Bedient, P. (2004). Assessing urban hydrologic prediction accuracy through event reconstruction. *Journal of Hydrology*, 299, 217–236.

Vogt, R., Christen, A., Rotach, M. W., Roth, M. & Satyanarayana, A. N. V. (2006). Temporal dynamics of CO_2 fluxes and profiles over a Central European city. *Theoretical and Applied Climatology*, 84, 117–126.

Wang, J., Endreny, A. & Nowak, D. J. (2008). Mechanistic simulation of tree effects in an urban water balance model. *Journal of the American Water Resources Association*, 44, 75–85.

Ward, H., Evans, J. & Grimmond, C. (2013). Multi-season eddy covariance observations of energy, water and carbon fluxes over a suburban area in Swindon, UK. *Atmospheric Chemistry and Physics*, 13, 4645–4666.

Xiao, Q., McPherson, E., Simpson, J. & Ustin, S. (2007). Hydrologic processes at the urban residential scale. *Hydrological Processes*, 21, 2174–2188.

5

ENVIRONMENTAL MEASUREMENTS IN BRIDGE CASE STUDIES

*Vicenzo Magliulo[1], Pierro Toscano[2], C.S.B. Grimmond[3],
Simone Kotthaus[3], Leena Järvi[4], Heikki Setälä[4],
Fredrik Lindberg[5], Roland Vogt[6], Tomasz Staszewski[7],
Anicenta Bubak[7], Afroditi Synnefa[8] and Mattheos Santamouris[8]*

[1]INSTITUTE FOR MEDITERRANEAN AGRICULTURAL AND FOREST SYSTEMS, [2]INSTITUTE OF BIOMETEOROLOGY,
[3]UNIVERSITY OF READING, [4]UNIVERSITY OF HELSINKI, [5]UNIVERSITY OF GÖTEBORG, [6]UNIVERSITY OF BASEL
[7]INSTITUTE FOR ECOLOGY OF INDUSTRIAL AREAS, AND [8]UNIVERSITY OF ATHENS

Introduction

'Urban metabolism' refers to the exchange and transformation of energy and matter between a city and its surroundings. Urban environments can be regarded as ecosystems (Pincetl et al. 2012) with their metabolism determined by the quantification of input and output flows. From a physical point of view, energy enters the system in the form of radiation and heat and influences local climate while being transported, transformed, stored and emitted. Quantitative information on the magnitude of energy flows is needed for management and environmental protection, while special emphasis is presently given to the exchange of carbon and water by the urban ecosystems, given their implications in global climate change issues.

A methodological framework to study urban metabolism was established in Chapter 4. In this chapter an observational approach to the study of urban metabolism across contrasting city environments is presented.

Core measurements carried out in BRIDGE case studies

The collection of meaningful data representative of a broad variety of conditions requires a careful choice of case studies, encompassing an extensive range of environmental and anthropogenic factors. In the BRIDGE project (Chrysoulakis et al. 2013), turbulent fluxes and significant components of urban flows were measured in five diverse European cities in terms of urban metabolism, and influenced by different policy and resource availability. Urban fluxes are strongly affected by the urban surface characteristics and modifications in land use (e.g. new buildings construction, increase of green areas etc.), notably at local and regional scales.

Five case studies were drawn from different parts of Europe, as presented in detail in Chapter 3: Helsinki, Athens, London, Firenze and Gliwice. Exchanges of energy, heat, moisture, carbon and pollutant were measured using different techniques, during the project, along with case study specific biophysical variables (Table 5.1). A common core of micrometeorological measurements was carried out in all cities except Athens, where it was not feasible to install an observation flux tower. These core measurements included: meteorological data, radiation budget and turbulent fluxes of latent heat, sensible heat and carbon dioxide (CO_2). The characteristics of each study site and the measurement activities undertaken are described in the rest of this section.

TABLE 5.1 Measurements carried out in BRIDGE case studies

	London	Athens	Firenze	Helsinki	Gliwice
Turbulent fluxes	✓		✓	✓	✓
Meteorological	✓	✓	✓	✓	✓
Air Quality (AQ)	✓	✓	✓	✓	✓
Urban vegetation	✓			✓	
Urban soils	✓			✓	✓
Stormwater/hydrology	✓			✓	✓
Dust turbulent fluxes			✓	✓	
Bio-monitoring of AQ			✓		
Indoor AQ		✓			
Urban heat island		✓			
Lidar dust profiling			✓		

London

In London the focus was the 'Central Activity Zone' (CAZ) (for details see Chapter 3). Two measurement towers were installed on the Strand Campus of King's College London (KCL, 51° 30′ N, 0° 7′ W), an area characterized as a high density urban zone for energy partitioning (Loridan & Grimmond 2012; Loridan et al. 2013) and as a compact midrise local climate zone (Stewart & Oke 2012; Kotthaus & Grimmond 2013a,b). The towers, approximately 60 m apart, had sensors at 49 m (King's Strand; KSS) and 39 m (King's Strand King's; KSK) above ground level which equates to about 2.2 and 1.9 times the mean building height. Full descriptions of the sites and the observations can be found in Kotthaus & Grimmond (2012, 2013a,b).

Three components of the surface energy balance were directly measured: net all-wave radiation and the turbulent fluxes of sensible and latent heat. The lower site (KSK) was established in 2008 and the other (KSS) in 2009. The radiation fluxes were measured with a net radiometer (CNR1 and CNR4, Kipp & Zonen); 15 minute averages are recorded. Cloud cover and height were determined from atmospheric backscatter measured with a ceilometer (CL31, Vaisala).

Turbulent latent and sensible heat fluxes were observed with Eddy Covariance (EC) systems consisting of a CSAT3 sonic anemometer (Campbell Scientific) and a Li7500 (Li7500A) open-path infrared gas analyser (LiCOR Biosciences) sampling at10 Hz. Fluxes were determined from 30 minute block averages. Automatic weather stations (WXT510 and WXT520, Vaisala) sampled air temperature, station pressure, horizontal wind speed and direction, and relative humidity at 5 second intervals. Precipitation from tipping bucket rain gauges (ARG100, Campbell Scientific) was totalled for 15 minute periods. The sonic anemometer, gas analyser, rain gauge and net radiometer were connected to Campbell Scientific CR3000 and CR5000 data loggers. Computers directly sampled the weather stations.

Additional measurements taken within the footprint of the EC flux towers included the water temperature of the River Thames obtained using Tinytag TG-4100 sensors at 0.2 m below the surface. The River Thames typically has a 6 m tidal range in the CAZ. In nearby parks (Embankment) and gardens (Middle Temple), soil temperature and moisture were measured using Delta-T Devices SM300 sensors buried approximately 0.1 m below the surface.

Natural Environment Research Council, Airborne Research & Survey Facility (NERC ARSF) flights occurred over Greater London during the study. This included a flight at an altitude of 900 m above ground level on 14 August 2008. On board, a 1064 nm wavelength Optech ALTM 3033 LiDAR system recorded first and last returns plus intensity values with a pulse frequency of 33 kHz using a maximum off-nadir view angle of 20°. The flight included a north to south transect area (approx. 0.65 km x 50 km) passing over the CAZ. This resulted in an on the ground mean point density of 0.71 m^{-2}. To avoid the large off-nadir angles at the edge of the field of view with a lower density points, 19 study areas were selected from the centre of the transect. The methods used to process the data are described in Lindberg & Grimmond (2011). From the data, surface cover information including vegetation characteristics were determined. Height attributes for each vegetation pixel were compared with a bare-earth Digital Surface Model (DSM) and those lower than 2.5 m above ground were removed from the data set. This means urban features such as building walls, power lines, masts etc. were still present and had to be removed in order to generate a vegetation canopy DSM. Vegetation has small gaps that allow laser light to penetrate and record additional returns at lower elevation. This was used to obtain a rough estimate of the crown base height needed to generate the trunk zone DSM.

Athens

This case study is focused on the municipality of Egaleo, a densely built urban area in the western part of Athens. The observations were performed at city, neighbourhood and building scale using different methodologies. Meteorological measurements were taken at the Thission meteorological station (37° 58′ N, 23° 46′ E) of the National Observatory of Athens. Hourly data for air temperature, relative humidity, wind speed and direction, precipitation, diffuse and total solar radiation, sunshine duration and air pressure were collected. Air quality data were retrieved from a network of stations installed in the greater Athens area measuring air pollution (sulfur dioxide (SO_2), nitrogen oxides (NO_x), carbon monoxide (CO), ozone (O_3), particulate matter (PM10, PM2.5), benzene (C_6H_6)) operated by the Directorate of Air Pollution and Noise of the Greek Ministry of Energy, Environment and Climate Change (Greek Ministry of Environment, Energy and Climate Change, 2010). The stations close to Egaleo were used for analysis. Traffic data, i.e. number of vehicles that passed through specific measurement locations (per hour), were also collected from the Ministry of Transport and Communication for 2009–2010.

In order to study the Urban Heat Island (UHI) phenomenon in the greater Athens area, a network of 17 fixed temperature stations was set up in different zones grouped into western, eastern, southern and northern zone stations according to their geographical location and thermal balances. In all stations the data were measured with fully calibrated high precision automatic miniature sensors (TinyTag), which were placed in white wooden boxes with lateral slots approximating the Stevenson screen to be protected for solar radiation and rain. Temperatures were measured at 15 minute intervals for 2009 and 2010. In order to assess the quality of the outdoor environment of the case study area the following measurements have been carried out using a mobile meteorological station and portable instrumentation:

- Concentrations of PM1, PM2.5, PM10 (Lighthouse 3016 IAQ Laser Particle Counter).
- Wind speed and direction by anemometers that have been placed at three different heights – 3.5, 7.5, 15.5 m – on the antenna of a mobile station.
- Air temperature, relative humidity (RH), air velocity and radiant temperature at a height of 1.5m.
- Measurements of the surface temperature of the urban fabric (building facades, roads and pavements) using an infrared thermometer (Cole Palmer) and camera (Thermovision 570).

Measurements were conducted during several days in the summer of 2009 between 10:00 and 17:30, in several locations in Egaleo area. The analysis of the results showed that increased surface and air temperatures and low wind speeds result in thermal discomfort for the people in the area.

In addition, an assessment of the impact of the outdoor environmental conditions (mainly thermal comfort and air pollution) on the indoor environment of residential buildings in the case study area was performed. The process followed is outlined below:

- selection and data collection of ten representative buildings of the Egaleo's building stock;
- distribution of questionnaires answered by the residents;
- indoor thermal comfort measurements: air temperature, relative humidity (Tiny Tag), air velocity (Dantec) and mean radiant temperature (INNOVA 1221);
- Measurements of PM1, PM2.5, PM10 concentrations (Model 8520 DUSTTRAKT).

Firenze

The observations were performed at city and neighbourhood scale, using different methodologies, while the core micrometeorological measurements exploited the existence of a pre-existing measurement tower operational when BRIDGE was started. All sensors were installed at the Ximeniano Observatory located in the city centre (43° 47′ N, 11° 15′ E) and were operated continuously for the duration of the project. The EC flux station was installed on a 3 m mast mounted on a typical tile roof of an ancient building of the Observatory. This roof is taller than the average surrounding buildings, free of obstacles, 7 m above the Observatory roof level and at 33 m above the street level (Matese et al. 2009; Gioli et al. 2012).

Turbulent fluxes of CO_2, momentum and sensible heat were collected, using a sonic anemometer and an open-path CO_2/water (H_2O) infrared gas analyser (Li7500). Raw data were acquired at the frequency of 20 Hz by Compact Eddy system (Matese et al. 2008), while flux data were computed at a 30 minute resolution, and quality checked with state-of-the-art procedures (Foken & Wichura 1996). Data were averaged over various timescales, to resolve daily, weekly, monthly and seasonal patterns of energy balance and surface. Averaging was undertaken only on the quality-checked data set. The tower was complemented with a new micrometeorological apparatus providing hourly measurements of particulate matter net flux by means of the EOLO system (Eddy cOvariance-based upLift Observation system, Fratini et al. 2007). This comprised a sonic anemometer (Metek USA-1) and a concentration system that includes an optical particle counter (OPC; CI-3100 series, Climet Instruments Co., Redlands, CA, USA) and a Multi-Channel Analyser (MCA8000, Amptek Inc., Bedford, MA, USA). This data was useful to compare with the outputs of the WRF-ACASA (Weather and Forecast model, Advanced-Canopy-Atmosphere-Soil Algorithm, Marras et al. 2011) simulation models for energy and mass fluxes in the city centre and for the dust deposition modelling framework (Tallis et al. 2011), to estimate the influence the proposed green space within the Parco San Donato would have on the local PM10 environment.

Meteorological observations were collected at the Ximeniano Observatory, an improvement of the pre-existing weather station, following the guidelines in the BRIDGE observation protocol. The average temperature recorded was 15.8 °C; an anomaly of +1.2 °C from the 1960–1991 baseline (warmest month July, mean temperature 26.3 °C and coldest month January, mean temperature 7.1 °C). Average annual rainfall was 934.9 mm; 24.2 mm above the climatological average.

An innovative microjoule LiDAR developed in cooperation with ENEA (the Italian national agency for new technologies, energy and sustainable economic development) was installed and tested on the Ximeniano Observatory. This resolved the first 5 km of atmosphere, providing measurements of horizontal and vertical profiles of dust as well as the height of the planetary boundary layer. A GRIMM aerosol spectrometer, capable of analysing the distribution of dust particles in 20 spectral bands in the range between 0.23 μm to 20 μm, was installed to complement remote observations of the LiDAR with detailed *in situ* information on the nature of the dust.

To better assess the quality of the outdoor environment measured above roof level and thus representative of mean air quality of the historic city centre, a 2BTECH 202 UV monitor and a UNITEC ETL2000 thick film sensor were installed for O_3 and CO concentrations respectively. Additional measurements taken within the footprint of the EC flux towers and in the neighbourhoods were provided by Tuscany Regional Agency for Environmental Protection (ARPAT) who measured concentrations of pollutants (PM10, CO_2, NO_x, SO_2, CO) for a network of five air quality 'traffic-oriented' monitoring stations including heavy and medium traffic and urban-background sites. Air quality was also monitored by means of suitable bio-indicators, providing a time-averaged picture of air quality at street level for the estimation of atmospheric trace metal deposition in the urban area of Firenze. At each site moss bags were exposed during three campaigns of measurement conducted during the periods March–April, May–July and August–October 2010. Two moss bags, used as control, were not exposed. After each campaign moss samples were analysed for arsenic (As), chromium (Cr), copper (Cu), iron (Fe), nickel (Ni), lead (Pb), vanadium (V) and zinc (Zn) by Inductively Coupled Plasma Atomic Emission Spectrometry.

Gliwice

The flux measurements in Gliwice were located 500 m west of the city centre on a building of the Silesian University of Technology on a balcony at the side of a gable roof (50° 17′ 38.01″ N, 18° 40′5 3.21″ E). The building height is approximately 25 m and an 8 m mast was installed on the balcony. The top of the mast was 3 m above the gable and the horizontal distance to the gable was 6 m. The line gable-to-mast pointed southwest. The effective height of the turbulent flux measurements above ground level was 29 m.

The set-up consisted of an EC system (sonic RM Young 81000V, LiCOR open-path gas analysers LI7500), a net radiometer (Kipp & Zonen CNR1) and sensors for air temperature and humidity (Vaisala HMP45). A data logger (Campbell CR1000) processed the data using the methods outlined in Aubinet et al. (2012). In addition the raw data (10 Hz) were kept for further analysis. In total 23 per cent of the data had to be discarded from flux calculations due to a range of conditions including rain/snow and maintenance. Monthly totals were calculated from monthly daily means to account for missing data. The measurements were carried out from the end of March 2010 until the end of February 2011.

Wind measurements were variable, thus three main wind directions were identified. As part of a weak diurnal wind regime night-time winds from the east were more frequent, while during the day-time winds were more often coming from the northwest and south-southwest. The month to month variation was dominated by synoptic conditions. The fetch of the approaching flow in these three sectors was typically around 1 km and consisted of built-up areas. Due to the relatively low measurement height the fluxes were affected by local-scale heterogeneities.

Characteristics of air quality were identified based on the continuous measurements carried out at the Voivodeship Inspectorate of Environmental Protection monitoring station in Gliwice. The station is located about 1 km from the Silesian University of Technology in Gliwice.

Helsinki

In Helsinki, most of the measurements carried out in the BRIDGE project have been made at the urban measurement station SMEAR III (Järvi et al. 2009a). The main site is the semi-urban Kumpula site located 5 km northeast of the centre of Helsinki. The meteorological variables, pollutant concentrations and turbulent fluxes of sensible and latent heat, CO_2 and aerosol particles have been measured continuously with the first observations in 2003.The fluxes are measured using the EC technique and the set-up consists of an ultrasonic anemometer (USA-1, Metek GmbH, Germany), open- and closed-path infrared gas analysers (LI-7000 and LI-7500), and a particle counter (WCPC, TSI-3781, TSI Incorporated, USA). During the project, the turbulent flux measurements were extended to the Helsinki centre at two locations (Hotel Torni and Erottaja fire station) where the surface cover is highly built-up (Nordbo et al. 2013).

The EC technique requires detailed planning both in terms of location and choice of instrumentation in urban areas. Typically either open-path infrared gas analysers, or closed-path analysers with long measurement tubes, are used. The open-path measurements are interfered with by the surface heating correction in the case of CO_2 emissions (Järvi et al. 2009b), whereas the long measurement tubes result in increased attenuation of water vapour as a function of relative humidity (Nordbo et al. 2012a).

In addition to the atmospheric measurements, the ecophysiology of urban trees was monitored in Viikki, 5 km northeast of Kumpula. The measurements of e.g. sap flow, soil temperature and CO_2 concentration were made in two streets where the trees were planted in 2002.

Storm water runoff was monitored from three water catchments located around Helsinki. The areas differ in their population density and fraction of impervious surfaces and are accordingly called low-density, medium-density and high-density catchments. In addition to water quantity, quality was also analysed, in terms of nutrients and potentially toxic elements.

Results

The richness of measurement protocols carried out yielded a variety of data sets suitable for use in BRIDGE in modelling activities, as well as in the development of the Decision Support System (DSS). A brief outline of the outcome of the common core micrometeorological measurements is reported here. All urban sites had shortcomings in terms of the representativeness of the flux observation for the whole area that the DSS needs to address. Thus, measured emissions must be considered in the context of the source area being measured, since the tower footprint covered only a fraction of the DSS areas of interest and the range of urban canopy types.

A more detailed analysis of energy and carbon fluxes might explain the different behaviour of the cities under investigation, by taking into account anthropogenically related variables, such as delivery rate of gas for heating or traffic intensity, the nature of the urban canopy and the presence of vegetation in the tower footprint. That analysis is beyond the scope of this chapter, but some of these issues been included in articles discussing the individual sites (Firenze: Gioli et al. 2012; Gliwice: Lesniok et al. 2010; Helsinki: Vesala et al. 2008; Järvi et al. 2009a,b,c; Nordbo et al. 2012a,b;; London: Kotthaus & Grimmond 2012, 2013a,b).

CO₂ fluxes

Turbulent exchanges of CO_2 (F_c) varied markedly across the sites, with differences in mean values spanning one order of magnitude (Figure 5.1), but with clear seasonal dynamics at each site. In all cities mean fluxes remained positive and cities acted as sources of CO_2.

Unlike terrestrial ecosystems, cities are dominated by anthropogenic CO_2 emissions so that the budget of this greenhouse gas is less driven by incoming solar energy. Rather, urban CO_2 emissions are mainly attributed to human activities (e.g. traffic, space heating) and photosynthesis is reduced to urban green spaces, which often cover smaller areas compared to the impervious surfaces. This leads to

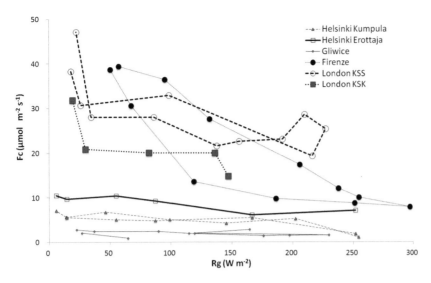

FIGURE 5.1   Monthly mean values of net carbon dioxide exchange (Fc) in BRIDGE case studies, as a function of incoming global radiation (R_g)

emphasizing the importance of land cover controls, urban planning (Nordbo et al. 2012b) and heating activities (Kotthaus & Grimmond 2012). Still, global radiation (R_g) can be used as a proxy for many anthropogenic activities related to carbon emissions (e.g. from combustion).

In all case study cities, monthly mean net CO_2 fluxes are highest in winter with maximum values around 40 µmol m^{-2} s^{-1} observed in London and Firenze and one month (December 2011) at the London KSS site reaching even 47 µmol m^{-2} s^{-1} (Figure 5.1). At Firenze, long-term CO_2 fluxes measured (Gioli et al. 2012, 2013) have always been a net source, with a small inter-annual variability associated with a high seasonality, ranging from 11.4 µmol m^{-2} s^{-1} during the periods' contributions from heating to 34.6 µmol m^{-2} s^{-1} during winter. This results in a strong, linear correlation between the flux and air temperature ($R^2 = 0.74$, $T_{air} < 288$ K). The relative contributions of road traffic and domestic heating to the observed emissions have been estimated through multivariate analysis combined with inventory data and emission proxies such as traffic counters and gas network flow rates. This analysis revealed that domestic heating accounts for more than 80 per cent of observed CO_2 fluxes in the winter and for about half the fluxes in summer.

Seasonal hysteresis was evident in the relation between global radiation and carbon fluxes for Firenze. The fluxes of the first six months of the year were higher than those recorded in the following months, for equivalent levels of radiation. In London, April and May 2011 had exceptionally persistent periods of high pressure weather conditions so that global radiation actually exceeded those recorded in June and July. Hence, the hysteresis pattern is less evident.

Northern cities showed a marked linear decrease in emissions from December through April ($R^2 = 0.72$ and 0.77, for Helsinki Kumpula and London KSS, respectively), while a steady decline from January through July was evident in Firenze ($R^2 = 0.98$).

The CO_2 emissions in the centre of Helsinki (Erottoja) were higher or of the same order of magnitude as the emissions measured downwind from a large road at the semi-urban Kumpula site (Nordbo et al. 2013). Seasonality plays an important role in both aerosol particle and CO_2 emissions. While aerosol particle emission factors increased with decreasing temperature (Ripamonti et al. 2013), the timing of the growing season was evident in CO_2 emissions resulting in a CO_2 sink of 10 µmol m^{-2} s^{-1} (Järvi et al. 2012). This indicates the importance of urban green areas especially in relatively green cities such as Helsinki. Besides CO_2 uptake, vegetation also plays a role as a nocturnal emitter via soil respiration and in Kumpula annual estimates for this are 550 and 645 g C m^{-2} in 2008 and 2009, respectively (Järvi et al. 2012).

Fluxes in Gliwice were mostly consistent year round – with a trend of increased emission activity starting in December and lasting until April (R^2=0.92). Thus they reflect typical features of urban areas. From October to April the monthly averaged diurnal patterns show a maximum of 6 to 4 μmol m^{-2}s^{-1} around midday, with nocturnal values around 1 to 2 μmol m^{-2}s^{-1}. From May to August, the midday values decrease, with some uptake in July and August. When weekends and weekdays were separately analysed, the influence of the traffic on CO_2 fluxes was clearly detected. During weekdays the fluxes were around 4 to 6 μmol m^{-2}s^{-1} higher than on weekends. It is even possible to detect the difference between Saturdays and Sundays, where the latter have the least traffic and show uptake values of around −3 to −4 μmol m^{-2}s^{-1}. Overall, however, the emissions dominate. The net monthly averages are all positive indicating that the area is a net source of CO_2. Based on monthly mean net fluxes, the estimated total annual flux is 2.8 kg CO_2 m^{-2} year^{-1}. This is a low value compared with other studies. The relation between mean monthly air temperature and total monthly CO_2 exchange shows a well behaved negative relation.

Aerosol particle fluxes

In Firenze and Helsinki aerosol particle fluxes were also measured. The aerosol particle fluxes in Firenze have a less pronounced seasonal trend compared to *Fc*. However, the PM2.5 fluxes exhibit a pronounced weekend decrease (−39 per cent) highlighting that the contribution of heating to particle emissions is relatively small compared to road traffic (Gioli et al. 2013). Similarly, in Helsinki, both the long term CO_2 and aerosol particle number emissions depend strongly on the amount of local traffic, indicating the importance of vehicles in urban metabolism (Järvi et al. 2009c, 2012).

Energy balance fluxes

All urban canopies partitioned most of the radiation load as sensible heat, with the exception of the semi-urban site of Helsinki Kumpula. This site, with a large amount of vegetation in one sector, had comparable values for sensible heat and latent heat fluxes. The case study sites show some clear differences in the magnitude of sensible heat in relation to the total incoming solar radiation (Figure 5.2). When looking at the linear relation between the two energy fluxes, the lowest increase of sensible heat with increasing solar radiation was found at the Helsinki Erottaja site (*H* proportional to 0.22 R_n) while the highest increase (0.29) was observed in Firenze and the other Helsinki site (Kumpula). Erottaja and the two London sites (KSK and KSS) reveal the largest sensible heat fluxes for times with low/no incoming solar energy. At these urbanized sites, night-time sensible heat fluxes can be significant due to storage heat fluxes and considerable anthropogenic contributions (Kotthaus & Grimmond 2013a). In Gliwice, the sensible heat flux still dominates the latent heat flux during the whole year, as is typical in urban areas, with a Bowen Ratio of around 2, reflecting the vegetation and water availability in Gliwice during the measurement period.

Similarly, the monthly mean latent heat flux with incoming solar radiation was compared for each site. In Helsinki Kumpula, there is more green space in the tower footprint than at the other sites so there is a larger slope for this site than the others (Figure 5.3). The area with the smallest amount of vegetation is in London, so at this site there is no indication of a clear relation with solar radiation. The source of moisture to support the latent heat flux is from rainfall, when the surface is wet (Kotthaus &

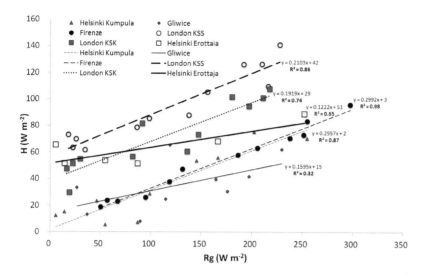

FIGURE 5.2 Monthly mean values of sensible heat flux in BRIDGE case studies, as a function of incident global radiation. Lines are linear regressions; R² values, as well as regression parameters, are also reported

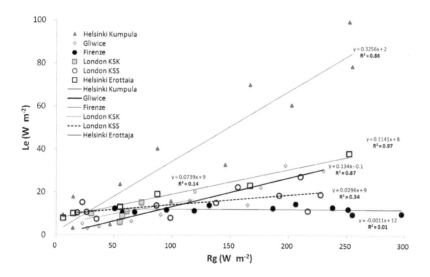

FIGURE 5.3 Monthly mean values of latent heat flux in BRIDGE case studies, as a function of incoming solar radiation. Lines are linear regressions; R2 values, as well as regression parameters, are also reported

Grimmond 2013a). As this occurs throughout the year and when there is less cloud present, the relation with solar radiation is not evident at the monthly timescale (Kotthaus & Grimmond 2013a,b). Similarly, in Firenze, the latent heat is very small (averaged only 10 W m⁻²) given the very small (less than 4 per cent) green area in the tower footprint.

TABLE 5.2 Mean and standard deviation values of summer and winter concentration of the main air pollutants in BRIDGE case studies (all data in $\mu g\ m^{-3}$)

	London		Athens		Firenze		Helsinki		Gliwice	
	summer	winter	summer	winter	summer	winter	summer	winter	summer	winter
PM 10	35.5 ± 6.4	41.6 ± 8.6	37.5 ± 2.9	45.7 ± 6.6	26.0 ± 1.8	41.7 ± 0.5	15.5	13.7	33.25 ± 4.0	88.5 ± 26.6
PM 2.5	20.7 ± 4.7	25.1 ± 7.2	32.2 ± 6.7	25.3 ± 3.3	14.7 ± 3.1	24.8 ± 2.7	8.58	10.3	18.75 ± 4.9	74.25 ±24.3
O₃	19.6 ± 2.9	12.4 ± 1.3	75 ± 12.3	39.2 ± 16.5	74.4 ± 6.4	27.3 ± 6.0	50.3 ± 27.1	45.8 ± 22.6	50.75 ± 13.6	22 ± 6.9
NOₓ	220.2 ± 47.4	252.9 ± 40.9	7.6 ± 3	31 ± 19	82.4 ± 16.3	158.4 ± 13.8	9.6 ± 10.8	18.8 ± 21.8	31.75 ± 7.4	69.5 ± 22.6
CO	0.7 ± 0.05	0.7 ± 0.08	0.5 ± 0.1	1.1 ± 0.9	0.52 ± 0.02	0.85 ± 0.04	–	–	1.15 ± 0.6	4.3 ± 1.1
SO₂	5.1 ± 1.5	7.45 ± 2.5	6.5 ± 2.2	13.3 ± 3.2	1.1 ± 0.1	1.6 ± 0.1	0.50 ± 0.13	0.79 ± 0.22	8 ±1.1	27 ± 9.5
Benzene	–	–	4.4 ± 1.2	6.1 ± 1.3	–	–	–	–	3.13 ± 1.1	7.7 ± 2.7

Air quality

The differences in urban fabric also influenced air quality, which varied markedly among cities (Table 5.2). London emissions are dominated by high levels of NO_x, while PM and CO were noticeably high in winter in Gliwice, whereas it is summertime O_3 in Firenze and Athens that is high. When analysing across sites, significant relationships between air quality traits and mean CO_2 emissions were evident. Fluxes were an inverse function of CO air concentration both in wintertime ($R^2 = 0.85$) and in summertime ($R^2 = 0.97$) and directly related to NO_x ($R^2 = 0.86$ and $R^2 = 0.97$, in wintertime and summertime, respectively). In Athens, PM10 and nitrogen dioxide (NO_2) exceeded limits at a number of measurement stations. O_3 concentration exceeded the alert threshold ($240\ \mu g\ m^{-3}$) during several days at stations close to Egaleo. These exceedences were due mostly to the high levels of sunshine and high temperatures that favour the formation of O_3.

The micro-scale moss bags trace metals concentrations, measured during different campaigns, have the expected highest concentrations for almost all elements at the high-traffic road sites, and lower values near sites with less traffic density in Firenze (Pellizzaro et al. 2013).

In Athens, particle number concentration measurements indoors were generally correlated with outdoor concentration characteristics in the absence of important indoor sources. Although concentrations outside the residences were quite high, for very airtight buildings concentrations were significantly lower (given the fact that no internal PM sources were found inside). In contrast, poor construction and high infiltration rates due to, for example, old wooden door and window frames resulted in high indoor concentrations. The presence of internal sources e.g. excessive smoking and cooking (e.g. frying) as expected resulted in high concentrations during the activity.

In the Gliwice case study, the high values for PM, SO_2, NO_x and CO in wintertime reflect the issue of 'low emissions' typical of the Silesian region, where the majority of isolated houses are heated by coal combustion with poor control of the emitted gases (Lesniok et al. 2010).

Indoor and outdoor thermal environment

The studies in Athens were carried out at a different scale with respect to the other cities, mainly addressing the outdoor and indoor microclimate. The case study area has high solar radiation levels and sunlight availability throughout the year and air temperatures range from 3 °C (February) to 40 °C (July). The yearly average value of relative humidity is about 60 per cent. Dominant wind direction is from the northeast with an average wind speed of 3 ms^{-1}. Precipitation levels are quite low, especially during the summer.

The UHI analysis showed that the case study area suffers from a strong UHI effect. Measurements between Egaleo and a suburban station indicate a mean UHI intensity for the monitoring period of 2–3 °C, reaching, however, in many occasions a difference of 6–8 °C. Modelling combined with measurements showed that the impact of increased outdoor temperatures in Egaleo compared to temperatures recorded in a suburban area (i.e. the UHI effect) results in an 74 per cent increase in cooling load. In addition the analysis of the results of the outdoor experimental campaign in the case study area showed that increased surface and air temperatures and low wind speeds result in thermal discomfort for the people in the area.

Measurements inside the selected buildings showed that indoor thermal discomfort is a serious problem. It was found that over 40 per cent of the maximum indoor temperatures are up to 35 °C, while 70 per cent of the mean indoor temperatures are up to 30 °C for the specific monitoring period. Indoor temperatures up to 38 °C have been recorded as well as hot spells of almost 21 consecutive hours over 34 °C. Comparative analysis of the occupants' responses received from questionnaires and the measured indoor conditions indicate that the thermal comfort perception of the users is in agreement with the air temperature measurements. Also, particle number concentration measurements indoors is generally correlated with outdoor concentration characteristics in the absence of important indoor sources. In addition, although concentrations outside the residences were quite high, for very airtight buildings, concentrations were significantly lower (given the fact that no internal PM sources were found inside). On the contrary, poor construction and high infiltration rates due to e.g. old wooden frames resulted in high indoor concentrations. The presence of internal sources, like excessive smoking and cooking (e.g. frying) as expected resulted in high concentrations during the activity.

Conclusions

Urban metabolic fluxes were observed at both local scale and micro-scale in BRIDGE case studies. They provided a composite picture of the various cities, with different seasonal characteristics of carbon, energy and pollutant exchanges. It was observed that CO_2 flux and air quality were significantly related, attributed to the importance of vehicle emissions for both. The measurements presented in this chapter created a valuable input data set in BRIDGE for both modelling exercise (see Chapters 7, 8, 9, 10 and 11 for details) and DSS development (see Chapter 16 for details).

References

Aubinet, M., Vesala, T. & Papale, D. (2012). *Eddy Covariance. A Practical Guide to Measurement and Data Analysis.* Springer, Heidelberg, p. 430.

Chrysoulakis, N., Lopes, M., San José, R., Grimmond, C.S.B., Jones, M.B., Magliulo, V., Klostermann, J.E.M., Synnefa, A., Mitraka, Z., Castro, E., González, A., Vogt, R., Vesala, T., Spano, D., Pigeon, G., Freer-Smith, P., Staszewski, T., Hodges, N., Mills, G. & Cartalis, C. (2013). Sustainable urban metabolism as a link between bio-physical sciences and urban planning: the BRIDGE project. *Landscape and Urban Planning*, 112, 100–117.

Foken, T. & Wichura, B. (1996). Tools for quality assessment of surface-based flux measurements. *Agricultural and Forest Meteorology*, 78(1–2): 83–105.

Fratini, G., Ciccioli, P., Febo, A., Forgione, A. & Valentini, R. (2007). Size-segregated fluxes of mineral dust from a desert area of northern China by Eddy Covariance. *Atmospheric Chemistry and Physics*, 7, 2839–2854.

Gioli, B., Toscano, P. Zaldei, A., Fratini, G. & Miglietta, F. (2013). CO2, CH4 and particles flux measurements in Florence, Italy. *Energy Procedia*, 40, 537–544.

Gioli, B., Toscano, P., Lugato, E., Matese, A., Miglietta, F., Zaldei, A. & Vaccari, F.P. (2012). Methane and carbon dioxide fluxes and source partitioning in urban areas: the case study of Florence, Italy. *Environmental Pollution*, 164, 125–131.

Greek Ministry of Environment, Energy and Climate Change (2010). *Annual Report on Atmospheric Pollution*, Directorate of Air Pollution and Noise. Available online at: http://www.ypeka.gr/LinkClick.aspx?fileticket= NIpIqbQgp4U%3d&tabid=490

Järvi, L., Hannuniemi, H., Hussein, T., Junninen, H., Aalto, P.P., Hillamo, R., Mäkelä, T., Keronen, P., Siivola, E., Vesala, T. & Kulmala, M. (2009a). The urban measurement station SMEAR III: continuous monitoring of air pollution and surface-atmosphere interactions in Helsinki, Finland. *Boreal Environmental Research*, 14 (Suppl. A), 86–109.

Järvi, L., Mammarella, I., Eugster, W., Ibrom, A., Siivola, E., Dellwik, E., Keronen, P., Burba, G. & Vesala, T. (2009b). Comparison of net CO_2 fluxes measured with open- and closed-path infrared gas analyzers in urban complex environment. *Boreal Environmental Research*, 14, 499–514.

Järvi, L., Rannik, Ü., Mammarella, I., Sogachev, A., Aalto, P.P., Keronen, P., Siivola, E., Kulmala, M. & Vesala, T. (2009c). Annual particle flux observations over a heterogeneous urban area. *Atmospheric Chemistry and Physics*, 9, 7847–7856.

Järvi, L., Nordbo, A., Junninen, H., Riikonen, A., Moilanen, J., Nikinmaa, E. & Vesala, T. (2012). Seasonal and annual variation of carbon dioxide surface fluxes in Helsinki, Finland, in 2006–2010. *Atmospheric Chemistry and Physics*, 12, 8475–8489.

Kotthaus S. & Grimmond, C.S.B. (2012). Identification of micro-scale anthropogenic CO_2, heat and moisture sources – processing Eddy Covariance fluxes for a dense urban environment. *Atmospheric Environment*, 57, 301–316.

Kotthaus S. & Grimmond, C.S.B. (2013a). Energy exchange in a dense urban environment – Part I: temporal variability of long-term observations in central London. *Urban Climate*, Available online. ISSN 2212–0955, http://dx.doi.org/10.1016/j.uclim.2013.10.002.

Kotthaus S. & Grimmond, C.S.B. (2013b). Energy exchange in a dense urban environment – Part II: impact of spatial heterogeneity of the surface. *Urban Climate*, Available online, ISSN 2212–0955, http://dx.doi.org/10.1016/j.uclim.2013.10.001.

Lesniok, M., Malarzewski, L. & Niedzwiedz, T. (2010). Classification of circulation types of Southern Poland with an application to air pollution concentration in Upper Silesia. *Physics and Chemistry of the Earth*, 35, 516–522.

Lindberg, F. & Grimmond, C.S.B. (2011). Nature of vegetation and building morphology characteristics across a city: influence on shadow patterns and mean radiant temperatures in London. *Urban Ecosystems,* 14, 617–634.

Loridan T. & Grimmond, C.S.B. (2012). Characterization of energy flux partitioning in urban environments: links with surface seasonal properties. *Journal of Applied Meteorology and Climatology*, 51, 219–241.

Loridan, T., Lindberg, F., Jorba, O., Kotthaus, S., Grossman-Clarke, S. & Grimmond, C.S.B. (2013). High resolution simulation of surface heat flux variability across central London with Urban Zones for Energy partitioning. *Boundary Layer Meteorology*, DOI: 10.1007/s10546–013–9797-y

Marras, S., Pyles, R.D. Sirca, C., Paw U, K.T., Snyder, R.L., Duce, P. & Spano, D. (2011). Evaluation of the Advanced Canopy-Atmosphere-Soil Algorithm (ACASA) model performance over Mediterranean maquis ecosystem. *Agricultural and Forest Meteorology*, 51, 730–745.

Matese, A., Gioli, B., Vaccari, F.P., Zaldei, A. & Miglietta, F. (2009). Carbon dioxide emissions of the city center of Firenze, Italy: measurement, evaluation, and source partitioning. *Journal of Applied Meteorology and Climatology*, 48, 1940–1947.

Matese, A., Alberti, G., Gioli, B., Toscano, P., Vaccari, F.P. & Zaldei, A. (2008). Compact(-)Eddy: a compact, low consumption remotely controlled Eddy Covariance logging system. *Computers and Electronics in Agriculture*, 64, 343–346.

Nordbo, A., Järvi, L. & Vesala, T. (2012a). Revised Eddy Covariance flux calculation methodologies – effect on urban energy balance. *Tellus B*, 48, 18184. http://dx.doi.org/10.3402/tellusb.v64i0.18184

Nordbo, A., Järvi, L., Haapanala, S., Wood, C. & Vesala, T. (2012b). Fraction of natural area as main predictor of net CO_2 emissions from cities. *Geophysical Research Letters*, 39, L20802, DOI: 10.1029/2012GL053087.

Nordbo, A., Järvi, L., Haapanala, S., Moilanen, J. & Vesala, T. (2013). Intra-city variation in urban morphology and turbulence structure in Helsinki, Finland. *Boundary-Layer Meteorology*, 146, 469–496.

Pellizzaro, G., Canu, A., Arca, A. & Duce, P. (2013). Heavy-metal biomonitoring by using moss bags in Florence urban area, Italy. EGU General Assembly 2013, held 7–12 April 2013 in Vienna, Austria, id. EGU2013–8506.

Pincetl, S., Bunje, P. & Holmes, T. (2012). An expanded urban metabolism method: toward a systems approach for assessing urban energy processes and causes. *Landscape and Urban Planning*, 107(3): 193–202.

Ripamonti, G., Järvi, L., Mølgaard, M., Hussein, T., Nordbo, A. & Hämeri, K. (2013). The effect of local sources on aerosol particle number size distribution, concentrations and fluxes in Helsinki, Finland. *Tellus B*, 65, 19786, http://dx.doi.org/10.3402/tellusb.v65i0.19786.

Stewart, I.D. & Oke, T.R. (2012). 'Local Climate Zones' for urban temperature studies. *Bulletin of the American Meteorological Society*, 120525055949004, doi:10.1175/BAMS-D-11–00019.1.

Tallis, M.J., Freer-Smith, P.H., Sinnett, D. & Taylor, G. (2011). Estimating the removal of atmospheric particulate pollution by the urban tree canopy of London, under current and future environments. *Landscape and Urban Planning*, 103, 129–138.

Vesala T., Järvi L., Launiainen S., Sogachev A., Rannik Ü., Mammarella I., Siivola E., Keronen P., Rinne J., Riikonen A. & Nikinmaa E. (2008). Surface-atmosphere interactions over complex urban terrain in Helsinki, Finland. *Tellus B*, 60, 188–199.

6

USE OF EARTH OBSERVATION TO SUPPORT URBAN MODELLING PARAMETERIZATION IN BRIDGE

Constantinos Cartalis[1], Marina Stathopoulou[1], Zina Mitraka[2] and Nektarios Chrysoulakis[2]

[1]UNIVERSITY OF ATHENS AND [2]FOUNDATION FOR RESEARCH AND TECHNOLOGY — HELLAS

Introduction

Urban studies conventionally use environmental and geographic information so as to depict the state of the urban environment, the characteristics of the urban landscape, the changes that have taken place in the course of time, the pressures on the urban area and the potential of varying techniques and technologies to ameliorate the thermal environment (Grimmond et al. 2010; Santamouris et al. 2011). Such information needs to be spatially and temporally dense, a fact which implies the need for monitoring networks or special experiments having the capacity to sufficiently reflect the area under investigation. Despite improvements in our understanding of urban processes and their impact on quality of life of city dwellers, few cities or urban agglomerations operate monitoring networks that may be considered sufficient.

As discussed in the next chapter, atmospheric models are increasingly employed to improve the understanding of processes that are related to local scale climate and air quality, the urban heat island and meso-scale circulations related to surface land cover and characteristics. Such processes are strongly influenced by the energy and momentum exchange between the atmosphere and the underlying surface; hence, the simulation of these processes depends on the accurate characterization of the surface properties. Information on surface physical properties (e.g. albedo, emissivity) and morphology (e.g. ground elevation, building height and geometry characteristics) is generally needed. Update-to-date information on these parameters can be provided by the current Earth Observation (EO) satellites and is expected to be improved in the near future. This chapter presents the potential of the current EO systems to support surface parameterization in model simulations that were performed in BRIDGE (Chrysoulakis et al. 2013a), and also providing some examples of such satellite derived products for the case study of Athens, Greece.

To many urban planners EO is providing a reliable alternative to conventional techniques (Esch et al. 2013). Yet considerable discussion is taking place on the real capacity of EO to support urban studies, including urban modelling. Such discussion mostly relates to the resolution (temporal, spatial and spectral) of satellite images. If the spatial resolution is low, studies may reflect inaccuracies, especially in the area of land use/cover changes. On the contrary, if the spatial resolution is high, the load of exploitable information from satellite data may highly exceed the respective load of ground data. In addition, temporal and spatial resolutions of a satellite image are anti-correlated, meaning that the better the spatial resolution, the worse the temporal one, and vice versa. When discussing satellite images, spectral resolution also needs to be explored, as the combined use of varying spectral channels may provide invaluable products for the state of the urban environment and the respective processes. For instance, although the study of many urban applications is based on images in the visible and near infrared parts of the spectrum, satellite images in the thermal infrared have an excellent potential in terms of supporting urban microclimatic studies.

In terms of urban modelling, different models are used, from meso-scale air quality models to urban canopy models. As discussed in more detail in Chapter 7, the modelling approach of BRIDGE was based on the cascade modelling technique (large to local scale) and includes the following steps: a) energy and water fluxes were measured and modelled at local scale; b) fluxes of carbon and pollutants were also modelled and their spatio-temporal distributions were estimated; c) fluxes were dynamically simulated in a 3D context by using state-of-the-art numerical models; and d) outputs of the above models resulted in indicators, which defined the state of the urban environment and were incorporated into the BRIDGE Decision Support System (DSS).

EO supported the models used in BRIDGE with detailed (spatially and temporally) information for such parameters as Land Surface Albedo (LSA), Land Surface Emissivity (LSE) and Land Surface Temperature (LST). It also provided topographic data in support of the development of Digital Elevation Models (DEMs), as well as estimates of the Aerosol Optical Thickness (AOT), i.e. a parameter considered essential for surface LSA calculations. Finally, it provided valuable information on land use/cover; such information, once provided in adequate resolution, allows the detection of changes which may be attributed to urban processes and patterns. To this end, the potential of EO is considered important, especially taking into consideration the developing wealth of satellite sensors and the overall improvement in their spatial, temporal and spectral resolutions. Below, a concise description regarding the methodology applied for each of the above parameters is given, along with some examples from the case study of Athens, as promoted within the BRIDGE project.

Satellite derived LSA

Satellite derived LSA products are used as inputs in meso-scale model simulations. LSA can be estimated from white-sky (completely diffuse) and black-sky (direct beam) albedo products as retrieved from Moderate Resolution Imaging Spectroradiometer (MODIS) observations for each elementary (dependent on the spatial resolution of the sensors) surface (Schaaf et al. 2002). It is noted that the spectral albedo is not a true surface property, but rather a characteristic of the coupled surface–atmosphere system. However, both black-sky and white-sky albedos are true surface properties and correspond to the limiting cases of point source and completely diffuse illumination. MODIS observations do not directly measure LSA; to this end spectral albedo is derived by directional integration of land surface reflectance recorded at the sensor, and is therefore dependent on the Bidirectional Reflectance Distribution Function (BRDF), which describes the dependency of reflectance on view and solar angles.

A directional sampling of surface reflectance from sensors such as MODIS can only be obtained by the accumulation of sequential observations over a specified time period. A 16-day period (or more) reflects a trade-off between the sufficiency of angular samples and the stability of surface reflectivity. These directional observations can be coupled with semi-empirical models to describe the BRDF and integrals necessary to provide spectral albedos. As discussed by Lucht et al. (2000), the BRDF can be expanded into a linear sum of terms (the so-called kernels), characterizing different scattering modes (isotropic, volumetric and geometric). The MODIS BRDF/albedo algorithm makes use of a kernel-driven, linear BRDF model which relies on the weighted sum of an isotropic parameter and two kernels (volumetric and radiometric) of viewing and illumination geometry to determine reflectance (Schaaf et al. 2002).

The black-sky albedo, as well as the white-sky albedo, are computed using polynomial expressions of the kernel weights, as described by Schaaf et al. (2002). The diffuse component can be expressed as a function of wavelength, optical depth, aerosol type and terrain contribution. Therefore, for partially diffuse illumination actually occurring, the spectral albedo may be approximated as a linear combination of the limiting cases. For this approximation, the fraction of diffuse radiation should be calculated; its calculation is straightforward as a function of solar zenith angle and AOT.

If Lambertian conditions are assumed, LSA may also be approximated from satellite images of high spatial resolution. Using the image data of Thematic Mapper (TM) of Landsat as recorded in the visible, near-infrared and mid-infrared spectral channels, the total shortwave albedo (0.25–5.1 μm), the visible albedo (0.4–0.7 μm) and the near-infrared albedo (0.7–5.0 μm) can be calculated at the spatial resolution of 30 m. In particular the processing technique includes:

- estimation of the incoming spectral radiative fluxes at the top of the atmosphere;
- atmospheric correction of the image data;
- correction of the incoming fluxes for the orientation of the surface and for the anisotropic reflection by the surface;
- conversion of the estimated surface leaving spectral radiance to surface spectral reflectance;
- conversion of the surface spectral reflectance to surface spectral albedo;
- conversion of the surface spectral albedo to broadband surface albedo as per Liang (2001).

For example, in the Athens case study, besides the MODIS derived LSA time series used in meso-scale modelling parameterization, LSA was also derived from a high-spatial resolution satellite image from Landsat 5 acquired over metropolitan Athens, under cloud free atmospheric conditions. The processing technique followed the above steps. The resulted shortwave albedo is shown in Plate 4. A zonal analysis of the product, leads to the conclusion that the most densely built areas of Athens exhibit a mean shortwave albedo value of 14.7 per cent, whereas all types of urbanized surfaces of the city can be represented by a mean value of 15.85 per cent (Stathopoulou et al. 2009).

Satellite derived LSE

Satellite derived LSE products are used in meso-scale models simulations. Daily emissivity maps (MODIS Level 2 emissivity product) are available as global maps at 1 km spatial resolution. The classification-based emissivity method proposed by Snyder et al. (1998) is used as developed with the linear BRDF models. Such models utilize spectral coefficients derived from laboratory measurements (Salisbury & D'Aria 1992, 1994: Salisbury et al. 1994: Snyder et al. 1997) of material samples. They also use structural parameters as derived from approximate descriptions of the cover type (Snyder & Wan 1998). LSE can be also be estimated with high resolution satellite data taking into account the 'mixed pixels' problem and the emissivity angular anisotropy (Mitraka et al. 2012).

Another methodology is exemplified in the following example: LSE was derived at spatial resolution of 30 m following the processing of three Landsat TM images acquired over metropolitan Athens during the warm season of the year. For each of these images, effective LSE in the 10–12 μm waveband was extracted applying the algorithm proposed by Caselles et al. (1991) and using a mean thermal emissivity value of 0.93 for the urbanized areas of Athens and a mean value of 0.98 for the vegetated surfaces (Stathopoulou & Cartalis 2007a). Then a composite LSE image was produced as a result of the overlay of LSE images and considering a mean LSE value for each pixel. In this way, a mean seasonal LSE image of Athens was produced (Plate 5), showing lower LSE values in the central and NW suburbs of Athens as compared to the N and NE ones, a fact which implies the presence of less vegetation and green open spaces in these areas.

LST and surface urban heat island

LST is a key parameter for mapping surface urban heat islands (SUHIs) and understanding the state of the thermal environment, including energy fluxes. It can be retrieved from thermal infrared image data at varying temporal and spatial resolutions (Prata et al. 1995: Sobrino & Jiménez-Muñoz 2005), with the principal preconditions for the retrieval being the correct estimation of the atmospheric effects and of LSE.

On the basis of LST, information about the SUHI characteristics of a city can be obtained, such as development and spatial pattern (heat island or heat sink), growth and evolution (SUHI area in km²), and intensity (LST difference observed between urban areas and the surrounding countryside).

In general, the processing steps for the estimation of LST and SUHI from satellite data are:

- Selection of satellite data and pre-processing (radiometric correction, cloud masking, geometric correction and projection to a standard geodetic system).
- Definition of methodology/technique for the analysis of thermal infrared image data to estimate LSE and LST. Special attention needs to be given, on the basis of land cover, to the distinction of urban to non-urban areas, as well as to the distinction of the urban areas as a function of the optical and thermal properties of commonly used building and paving materials.
- Application of image processing methodologies/techniques to satellite imagery with the final stage being the classification of the thermal surface pattern of the urbanized centres and the development of thermal maps of the urban areas.
- Application (potentially) of remote sensing downscaling techniques for low resolution data for selected cases.
- Use of high resolution satellite data in the identification of 'hot spot' events within the urban areas under study.
- Estimation of LST gradients within the urban areas and detection/measurement of the SUHI, including its extent and intensity.
- Link of LST/SUHI to land use/cover, physical interpretation and statistical analysis.
- Validation of results based on error analysis and determination of classification accuracies.

Many SUHI studies focus on the use of polar orbiting satellites that have adequate spatial resolution (~1 km) and daily revisiting capabilities, such as MODIS. For example, Hung et al. (2006) used MODIS data to perform a comparative study of the SUHIs of eight megacities in Asia. Besides MODIS data, they also utilized high spatial resolution Landsat satellite images to derive land cover information and to relate SUHI patterns to surface characteristics. The researchers adopted the Gaussian method proposed by Streutker (2002) and measured the spatial extents and intensities of the SUHIs of the eight megacities. The diurnal and seasonal patterns of the SUHIs revealed that all cities exhibited significant heat islands. Likewise, many researchers (Dousset & Gourmelon 2003; Stathopoulou et al. 2004; Tomlinson et al. 2012) have demonstrated that the use of low resolution data in large urban agglomerations may provide solid information on the presence, extension and differentiation of LST and SUHI. In practice, data from satellites with high revisit time such as MODIS or AVHRR, support the monitoring of the SUHI intensities at a frequent basis as well as the retrieval and mapping of the thermal inertia of the urban landscape; the latter supports the understanding of the roles that individual components of the city (such as parks, industrial complexes, etc.) play in the thermal patterns of the city (Voogt & Oke 1998). This requirement is particularly important if the link between SUHI intensity and air pollution is to be explored (Lo & Quattrochi 2003), or if satellite-derived LST is used to calculate energy parameters, such as the cooling degree days (Stathopoulou et al. 2006) or bioclimatic parameters, such as the thermal discomfort index (Stathopoulou et al. 2005).

LST can be also estimated with the use of satellite imagery of spatial resolutions from 60, to 90 to 120 m. Such imagery may be highly beneficial to detect 'hot spots' within urban areas or to examine in better spatial detail findings from 1 km resolution satellites. This requirement is of particular importance if the study area reflects a city of moderate size (Stathopoulou et al. 2004), or a region of particular interest within a city (Nichol 1996). At the same time, such imagery faces the problem of limited temporal resolution (revisit time of the order of 16 days), a fact which narrows the operational character of the application. It should be mentioned that the potential of downscaling techniques as applied to low spatial resolution satellite data (Stathopoulou & Cartalis 2009) has already been examined, with promising results in terms of the capacity of the methodology developed to adequately reflect the radiometric information at an improved spatial resolution. Jiménez-Muñoz & Sobrino (2003) have developed a generalized split-window algorithm for the retrieval of LST that can be applied to all thermal sensors characterized with a Full-Width Half-Maximum (FWHM) of around 1 μm. The main advantage of this algorithm is that it can be applied to different thermal sensors using the same equation and coefficients.

An LST estimation methodology is exemplified in the following example, where summertime Landsat images over the metropolitan Athens area were analysed: following the calibration process of the thermal channel image data, the generalized single-channel algorithm (Jiménez-Muñoz & Sobrino 2003) was applied. The algorithm requires knowledge of LSE for the application area, as well as the atmospheric water vapour content for the observation day. Therefore, the algorithm was applied using LSE values by land cover (Stathopoulou & Cartalis 2007b) and a mean value of water vapour content per month as measured from radiosonde data in Athens (Chrysoulakis & Cartalis, 2002). As shown in Figure 6.1, the mid-morning summertime LST spatial profile demonstrates the development of a negative SUHI (heat sink), with the extended dry surfaces of bare soil and low vegetation in the surrounding rural areas presenting higher LST values than the urbanized surfaces of the Athens basin. The latter, as characterized by lower sky view factor and higher thermal capacity, exhibits lower warming rate from incident solar radiation compared to rural areas at the periphery of the city (Stathopoulou & Cartalis 2007a).

Land use/cover

In order to determine the intensive interactions between humans and the environment in the cities, especially in the metropolitan areas and their peripheries, as well as to determine and monitor the main characteristics of change, Land Use/Land Cover (LULC) monitoring and extracting LULC type is important (Qian 2007). Urban land cover units can be distinguished by their characteristic pattern of built and open spaces, which form the background matrix of the city, whereas the different materials can be distinguished by their characteristic spectral response (Chrysoulakis et al. 2013b; Herold et al. 2003; Pauleit & Duhme 2000). Land cover units result from past and present human activities and form relatively stable features lasting in time. Furthermore, land cover units are the product of land use and development plans. Delineating these units leads to an assessment of past and current city planning decisions. In addition, planning concepts such as mixed housing areas in closed block buildings versus housing in multi-story high rise buildings can be evaluated. Land cover units also have a characteristic land use associated with them (Weber et al. 2005; Chorianopoulos et al. 2010).

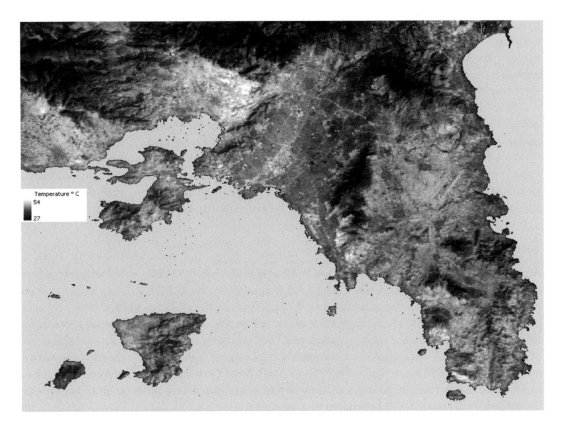

FIGURE 6.1 Surface temperature in Athens (daytime analysis) based on Landsat observations and auxiliary spatial data

Several classification and segmentation methods have been developed for the processing of satellite data for land cover/land use applications. Although supervised classification can be used for LULC classification purposes, the 'mixed pixel' problems can affect the results because of the nature of the classificatory criteria (Enderle & Weih 2005; Bhaskaran et al. 2010). The object-oriented classification method can be more successful than the pixel-based methods (Whiteside & Ahmad 2005; Chena et al. 2009; Chrysoulakis et al. 2013b) as it creates segments from image pixels which includes homogeneous, spatially contiguous regions.

The object-based classification process of satellite images is based on object-oriented software. Object-oriented software requires first segments which have to be created by pre-set homogeneity criteria. The scale, colour and form parameters are used to create homogeneous segments from pixels. These parameters are defined by the user in an empirical manner, whereas the scale parameter affects the heterogeneity of the pixels. The colour parameter defines balance between the homogeneity of a segment's colour and the homogeneity of its shape. The smoothness of an object's border with its compactness can be balanced with the form parameter. The segmentation process has been explained in detail by Baatz & Schäpe (2000). To achieve expected results, scale factor and other segmentation criteria are chosen empirically and interactively to demonstrate the morphology of the objects (Ranasinghe 2008).

Satellite images of high spatial resolution are required to map the LULC of the city giving emphasis to differentiating between urban and non-urban uses: 1) high-density urban use; 2) low-density urban use; 3) cultivated/exposed land; 4) cropland or grassland; 5) forest; and 6) water. The LULC change analysis is performed on the basis of statistics (coverage percentage) extracted from the produced land use/cover maps of the city area.

The methodology used to map the dynamics of urban growth in a study area includes pre-processing of high resolution images before urban feature extraction and the use of orthorectified products as reference data sets to geometrically correct the images. The brightness normalization method proposed by Wu (2004) can be applied to handle difficulties in quantifying urban composition. Although urban components show significant spectral differences (e.g. dark and bright soil), they share common characteristics. The normalization technique is used to highlight the shape information, while minimizing the effects of absolute reflectance values, by reducing the spectral variations between pure land use types.

Satellite derived DEMs

Advanced Spaceborne Thermal Emission and Reflection Radiometer (ASTER) stereo imagery has been used in BRIDGE for DEM generation at city and regional scales, for all BRIDGE case studies. For the production of a DEM from optical satellite data, the respective satellite sensors should have stereo coverage capabilities (Toutin 2001). The methods for applying stereoscopy to push-broom scanners generally use the rigorous photogrammetric solution (collinearity and coplanarity conditions for the conic perspective of a single image line), but also take into account the displacement of the satellite (cylindrical perspective) to link the equations between themselves (Toutin 2004). Since the parameters of neighbouring lines are highly correlated and satellite positions and attitude can be computed from on-board recording systems, the mathematical equations can be reduced to a minimum of eight to ten unknowns, depending on the mathematical development and the algorithm implementation of the solution. A stereo matching process to extract elevation parallaxes is applied (Toutin 2008). In particular, the basic characteristics of stereoscopy and its application to the ASTER for DEM generation have been recently reviewed by Toutin (2008). ASTER consists of three separate instruments subsystems, each operating in a different spectral region, using separate optical system. These subsystems are the Visible-Near Infrared (VNIR), the Short-Wave Infrared (SWIR) and the Thermal Infrared (TIR). The spatial resolution varies with wavelength: 15 m in the VNIR, 30 m in the SWIR and 90 m in the TIR. The VNIR subsystem consists of two telescopes – one nadir looking with a three band detector (Channels 1, 2 and 3N) and the other backward looking (27.7° off-nadir) with a single band detector (Fujisada 1994; Abrams 2000). The data products provided by the ASTER have been summarized by Yamaguchi et al. (1998).

The vertical accuracy of a DEM product is assessed using Global Positioning System (GPS) measurements. Elevation accuracy is generally of the order of the system instantaneous field of view (or approximately the pixel spacing), also taking into account the Base to Height ratio (B/H). The theoretical accuracy of ASTER DEMs is therefore governed by the accuracy of the control data, the B/H and the matching. Since an error of ± 0.5–1 pixel or better for the parallax measurements in the automated matching process has been achieved with different data sets from other sensors, the potential relative accuracy for the elevation with the ASTER stereo data (B/H = 0.6, pixel spacing of 15 m) can be of the order of 12–25 m or better depending of the type of terrain. In addition, the accuracy of the DEM will also depend on the geometric parameter calibration, as well as the accuracy of the ephemeris and attitude data for computing the direct georeferencing.

The generation of a DEM for the broader Athens area using ASTER stereo-pairs is now briefly described: ASTER Level 1A data was projected to the map of Athens using the radiometric calibration and geometric correction coefficients for resampling (Fujisada 1998). Auxiliary and validation data sets for the study area include contour lines and shoreline digitized from 1:50.000 topographic maps, GPS measurements and the road network. The methodology for the production of DEM as described in Chrysoulakis et al. (2004a, b) and Nikolakopoulos et al. (2006) was applied to extract a high resolution (15 m pixel) ASTER DEM that was further used to orthorectify the respective ASTER multispectral imagery.

Next, virtual images were created using the respective VNIR and SWIR channels, as well as the extracted DEM. As a result, multispectral orthorectified images were created, each containing the respective DEM as a separate pseudochannel. In practice two mosaics were initially created: one for the DEM pseudochannels and one for the VNIR and SWIR channels of each image. In the former case, the radiometric values of each DEM were retained and a bilinear interpolation resampling method is used. In the latter case, the colour balancing and histogram matching functions were used and the radiometric values of VNIR and SWIR channels are slightly modified, ensuring the colour homogeneity of all scenes. The two mosaics were finally merged to one master mosaic covering the broader area of Athens. This mosaic was finally fine-tuned using the road network vectors of the area. In this way the plannimetric (xy) error was corrected to ± 0.5 pixel root mean square deviation (RMSE) using a 2D transformation, resulting in an planimetric accuracy of the corrected DEM better than 15 m. The produced DEM is shown in Plate 6.

Conclusions

A main conclusion is that EO is considered a valuable tool for urban modelling, as it can provide valuable spatial and temporal information regarding the topography and the LULC types of the study areas, as well as of such parameters as LSA, LSE, LST and AOT, used to parameterize atmospheric models at regional and local scales in BRIDGE. The combined use of satellite data in various spectral regions is also beneficial as the derived urban characteristics (land use/cover, vegetation cover) provide considerable insights with respect to the presence, the intensity and variability of the SUHI. Results can be further improved from the combined use of ground data and can also be used as the starting point for the estimation of such products as the thermal comfort index, or the cooling degree days.

Disadvantages do exist and relate mostly to the spatial and temporal resolutions of satellite images as well as to the knowledge of the properties of the emitting surfaces. In particular, urban studies which need satellite images of high temporal resolutions (order of hours), correspond to low spatial resolutions (order of 1 km), a fact which may be considered highly confining for detailed results (Stathopoulou & Cartalis 2007b). In the event that satellite images of medium resolution are used instead (120–250 m), the respective temporal resolution drops considerably (from 2–3 days to 16 days). However, given that urban studies focus on phenomena that in the long run reflect timescales exceeding those of hours or even days, such (temporal and spatial) resolutions may be considered adequate for large urban agglomerations. Problems do arise, however, in smaller urban areas as low or medium spatial resolutions may lead to only a few pixels of exploitable information. Problems also arise in the examination of the daily variation of microclimatic parameters, as limited information of high spatial resolution is available from satellite night passes. The latter may be considered the most pronounced drawback in terms of existing satellite sensors/missions, especially if the scope of study reflects the state of the thermal environment.

Overall the potential of EO for urban modelling parameterization is considered highly important, especially taking into consideration the wealth of existing satellite sensors and the forthcoming satellite missions (e.g. Sentinels) combined with the lack of extended and adequately operated ground monitoring networks. The potential is further enhanced in the event of cities of large size or cities falling in the category of 'mega cities'.

References

Abrams, M. (2000). ASTER: data products for the high spatial resolution imager on NASA's EOS-AM1 platform. *International Journal of Remote Sensing*, 21, 847–861.

Baatz, M., & Schäpe, A. (2000). Multiresolution Segmentation: an Optimization Approach for High Quality Multi-Scale Image Segmentation. In: Strobl, J., Blaschke, T., Griesebener, G. (Eds.), Angewandte Geographische Informationsverarbeitung XII. Beiträge zum AGIT-Symposium (Salzburg, Austria), Herbert Wichmann Verlag, Kahrlsruhe, pp. 12–23.

Bhaskaran, S., Paramananda, S., & Ramnarayan, M. (2010). Per-pixel and object-oriented classification methods for mapping urban features using Ikonos satellite data. *Applied Geography*, 30, 650–665.

Caselles, V., Lopez Garcia, M. J., Melia, J., & Perez Cueva, A. J. (1991). Analysis of the heat island effect of the city of Valencia, Spain through air temperature transits and NOAA satellite data. *Theoretical and Applied Climatology*, 43, 195–203.

Chena, M., Sua, W., Lia, L., Zhanga, C., Yuea, A., & Lia, H. (2009). Comparison of pixel-based and object-oriented knowledge-based classification methods using SPOT5 imagery. *WSEAS Transactions on Information Science and Applications*, 3, 477–489.

Chorianopoulos I., Pagonis T., Koukoulas S., & Drymoniti S. (2010). Planning, competitiveness and sprawl in the Mediterranean city: the case of Athens, *Cities*, 27, 249–259.

Chrysoulakis, N., Lopes, M., San José, R., Grimmond, C. S. B., Jones, M. B., Magliulo, V., Klostermann, J. E. M., Synnefa, A., Mitraka, Z., Castro, E., González, A., Vogt, R., Vesala, T., Spano, D., Pigeon, G., Freer-Smith, P., Staszewski, T., Hodges, N., Mills, G., & Cartalis, C. (2013a). Sustainable urban metabolism as a link between bio-physical sciences and urban planning: the BRIDGE project. *Landscape and Urban Planning*, 112, 100–117.

Chrysoulakis, N., Mitraka, M., Stathopoulou, M., & Cartalis, C. (2013b). A comparative analysis of the urban web of the greater Athens agglomeration for the last 20 years period on the basis of landsat imagery. *Fresenius Environmental Bulletin*, 22, 2139–2144.

Chrysoulakis, N., Abrams, M., Feidas, H., & Velianitis, D. (2004a). Analysis of ASTER Multispectral Stereo Imagery to Produce DEM and Land Cover Databases for Greek Islands: The REALDEMS Project. In: Prastacos, P., Cortes, U., De Leon, J. L., Murillo, M. (Eds), Proceedings of e-Environment: Progress and Challenge, 411–424.

Chrysoulakis, N., Diamandakis, M., & Prastacos, P. (2004b). GIS Based Estimation and Mapping of Local Level Daily Irradiation on Inclined Surfaces. In: Toppen, F. and P. Prastacos (Eds), Proceedings of the 7th AGILE Conference on Geographic Information Science, 587–597.

Chrysoulakis, N., & Cartalis, C. (2002). Improving the estimation of land surface temperature for the region of Greece: adjustment of a split window algorithm to account for the distribution of precipitable water. *International Journal of Remote Sensing*, 23, 871–880.

Dousset, B., & Gourmelon, F. (2003). Satellite multi-sensor data analysis of urban surface temperatures and landcover. *Journal of Photogrammetry and Remote Sensing*, 58, 43–54.

Enderle, D., & Weih, R. (2005). Integrating supervised and unsupervised classification methods to develop a more accurate land cover classification. *Journal of the Arkansas Academy of Science*, 59, 65–73.

Esch, T., Taubenböck, H., Chrysoulakis, N., Düzgün, H. S., Tal, A., Feigenwinter, C., & Parlow, E. (2013). Exploiting Earth Observation in Sustainable Urban Planning and Management – the GEOURBAN Project. In: Proceeding of Joint Urban Remote Sensing Event JURSE 2013, Sao Paulo, Brazil, April 21–23.

Fujisada, H. (1998). ASTER level-1 data processing algorithm. *IEEE Transactions on Geoscience and Remote Sensing*, 36, 1101–1112.

Fujisada, H. (1994). Overview of ASTER instrument on EOS-AM1 platform. In: Proceedings of SPIE, 2268, 14–36.

Grimmond, C. S. B., Roth, M., Oke, T. R., Au, Y. C., Best, M., Bettse, R., Carmichael, G., Cleugh, H., Dabberdt, W., Emmanuel, R., Freitas, E., Fortuniak, K., Hannal, S., Kleinm, P., Kalkstein, L. S., Liu, C. H., Nickson, A., Pearlmutter, D., Sailor, D., and Voogt, J. (2010). Climate and more sustainable cities: climate information for improved planning and management of cities (producers/capabilities perspective). *Procedia Environmental Sciences*, 1, 247–274.

Herold, M., Gardner, M., & Roberts, D. A. (2003). Spectral resolution requirements for mapping urban areas. *IEEE Transactions on Geoscience and Remote Sensing*, 41, 1907–1919.

Hung, T., Uchihama, D., Ochi, S., & Yasuoka, Y. (2006). Assessment with satellite data of the urban heat island effects in Asian mega cities. *International Journal of Applied Earth Observation and Geoinformation*, 8, 34–48.

Jiménez-Muñoz, J. C. & Sobrino, J. A. (2003). A generalized single-channel method for retrieving land surface temperature from remote sensing data. *Journal of Geophysical Research*, 108 (D22), 4688, doi:10.1029/2003JD003480

Liang, S. (2001). Narrowband to broadband conversions of land surface albedo: I. Formulae. *Remote Sensing for Environmental Sciences*, 76, 213–238.

Lo, C. P., & Quattrochi, D. A. (2003). Land-use and land-cover change, urban heat island phenomenon, and health implications: a remote sensing approach. *Photogrammetric Engineering and Remote Sensing*, 69, 1053–1063.

Lucht, W., Schaaf, C. B., & Strahler, A. (2000). An algorithm for the retrieval of albedo from space using semiempirical BRDF models. *IEEE Transactions on Geoscience and Remote Sensing*, 38, 977–998.

Mitraka, Z., Chrysoulakis, N., Kamarianakis, Y., Partsinevelos, P., & Tsouchlaraki, A. (2012). Improving the estimation of urban surface emissivity based on sub-pixel classification of high resolution satellite imagery. *Remote Sensing of Environment*, 117, 125–134.

Nichol, J. E. (1996). High-resolution surface temperature patterns related to urban morphology in a tropical city: a satellite-based study. *Journal of Applied Meteorology*, 35, 135–146.

Nikolakopoulos, K., Kamaratakis, E., & Chrysoulakis, N. (2006). SRTM vs ASTER elevation products. Comparison for two regions in Crete, Greece. *International Journal of Remote Sensing*, 27, 4819–4838.

Pauleit, S., & Duhme, F. (2000). Assessing the environmental performance of land cover types for urban planning. *Landscape and Urban Planning*, 52, 1–20.

Prata, A., Caselles, V., Coll, C., Ottle, C., & Sorbino, J. (1995). Thermal remote sensing of land surface temperature from satellites: current status and future prospects. *Remote Sensing Reviews*, 12, 175–224.

Qian, J.-H. (2007). Application of a variable-resolution stretched grid to a regional atmospheric model with physics parameterization. *Advances in Earth Science*, 22, 1185–1190.

Ranasinghe, A. (2008). Multi scale segmentation techniques in object oriented image analysis. Asian Conference on Remote Sensing ACRS 2008, Colombo, Sri Lanka.

Salisbury, J. W., D'Aria, D. M., & Wald, A. (1994). Measurements of thermal infrared spectral reflectance of frost, snow, and ice. *Journal of Geophysics Research*, 99, 24,235–24,240.

Salisbury, J. W., & D'Aria, D. M. (1994). Emissivity of terrestrial materials in the 3–5 μm atmospheric window. *Remote Sensing of Environment*, 47, 345–361.

Salisbury, W., & D' Aria, D. M. (1992). Emissivity of terrestrial materials in the 8–14 mm atmospheric window. *Remote Sensing of Environment*, 42, 83–106.

Santamouris, M., Synnefa, A., & Karlessi, T. (2011). Using advanced cool materials in the urban built environment to mitigate heat islands and improve thermal conditions. *Solar Energy*, 85, 3085–3102.

Schaaf, C. B., Gao, F., Strahler, A. H., Lucht, W., Li, X. W., Tsang, T., Strugnell, N. C., Zhang, X. Y., Jin, Y. F., Muller, J. P., Lewis, P., Barnsley, M., Hobson, P., Disney, M., Roberts, G., Dunderdale, M., Doll, C., d'Entremont, R. P., Hu, B. X., Liang, S. L., Privette, J. L., & Roy, D. (2002). First operational BRDF, albedo nadir reflectance products from MODIS. *Remote Sensing of Environment*, 83, 135–148.

Snyder, W. C., Wan, Z., Zhang, Y., & Feng, Y.-Z. (1998). Classification-based emissivity for land surface temperature measurement from space. *International Journal of Remote Sensing*, 19, 2753–2774.

Snyder, W. C., & Wan, Z., (1998). BRDF models to predict spectral reflectance and emissivity in the thermal infrared. *IEEE Transactions on Geoscience and Remote Sensing*, 36, 214–225.

Snyder, W. C., Wan, Z., Zhang, Y., & Feng, Y. (1997). Thermal infrared (3±14 μm) bidirectional reflectance measurements of sands and soils. *Remote Sensing of Environment*, 60, 101–109.

Sobrino, J. A., and Jiménez-Muñoz, J. C. (2005). Land surface temperature retrieval from thermal IR data: an assessment in the context of the Surface Processes and Ecosystem Changes through Response Analysis (SPECTRA) mission. *Journal of Geophysical Research*, 110, D16103, doi:10.1029/2004JD005588.

Stathopoulou, M., Synnefa, A., Cartalis, C., Santamouris, M., Karlessi, T., & Akbari, H. (2009). A surface heat island study of Athens using high-resolution satellite imagery and measurements of the optical and thermal properties of commonly used building and paving materials. *International Journal of Sustainable Energy*, 28, 59–76.

Stathopoulou, M., & Cartalis, C. (2009). Downscaling AVHRR land surface temperatures for improved surface urban heat island intensity estimation. *Remote Sensing of Environment*, 113, 2592–2605.

Stathopoulou, M., & Cartalis, C. (2007a). Study of the Urban Heat Island of Athens, Greece During Daytime and Night-time. In: IEEE Conference Proceedings of the 2007 Urban Remote Sensing Joint Event, Paris, France, 11–13 April 2007 (pp. 1–7).

Stathopoulou M., & Cartalis C. (2007b). Daytime urban heat island from Landsat ETM+ and Corine land cover. *Solar Energy*, 81, 358–368.

Stathopoulou, M., Cartalis, C., & Chrysoulakis, N. (2006). Using midday surface temperature to estimate cooling degree-days from NOAA-AVHRR thermal infrared data: an application for Athens, Greece. *Solar Energy*, 80, 414–422.

Stathopoulou, M., Cartalis, C., Keramitsoglou, I., & Santamouris, M. (2005). Thermal Remote Sensing of Thom's Discomfort Index (DI): Comparison with *In Situ* Measurements. In: Proceedings of the Remote Sensing 2005 Symposium (SPIE 2005), Brugge, Belgium, 19 September 2005.

Stathopoulou, M., Cartalis, C., & Keramitsoglou, I. (2004). Mapping micro-urban heat islands using NOAA/AVHRR images and CORINE land cover: an application to coastal cities of Greece. *International Journal of Remote Sensing*, 25, 2301–2316.

Streutker, D. R. (2002). A remote sensing study of the urban heat island of Houston, Texas. *International Journal of Remote Sensing*, 23, 2595–2608.

Tomlinson, C. J., Chapman, L., Thornes, J. E., & Baker, C. J. (2012). Derivation of Birminghams summer surface urban heat island from MODIS satellite images. *International Journal of Climatology*, 32, 214–224.

Toutin, T. (2008). ASTER DEMs for geomatic and geoscientific applications: a review. Elevation Modelling from Satellite VIR Data: A Review. *International Journal of Remote Sensing*, 29, 1855–1875.

Toutin, T. (2004). Geometric processing of remote sensing images: models, algorithms and method. *International Journal of Remote Sensing*, 25, 1893–1924.

Toutin, T. (2001). Elevation Modelling from Satellite VIR Data: A Review. *International Journal of Remote Sensing*, 22, 1097–1125.

Voogt, J. A., & Oke, T. R. (1998). Effects of urban surface geometry on remotely-sensed surface temperature. *International Journal of Remote Sensing*, 19, 895–920.

Weber, C., Petropoulou, C., & Hirsch, J. (2005). Urban development in the Athens metropolitan area using remote sensing data with supervised analysis and GIS. *International Journal of Remote Sensing*, 26, 785–796.

Whiteside, T., & Ahmad, W. (2005). A Comparison of Object-Oriented and Pixel-Based Classification Methods for Mapping Land Cover in Northern Australia. In: Proceedings of SSC2005 Spatial Intelligence, Innovation and Praxis: The National Biennial Conference of the Spatial Sciences Institute, Melbourne, Australia, pp. 1225–1231.

Wu, C., (2004). Normalized spectral mixture analysis for monitoring urban composition using ETM+ imagery. *Remote Sensing of Environment*, 93, 480–492.

Yamaguchi, Y., Kahle, A., Tsu, H., Kawakami, T., & Pniel, M. (1998). Overview of advanced spaceborne thermal emission and reflection radiometer (ASTER). *IEEE Transactions on Geosciences and Remote Sensing*, 36, 1062–1071.

7

MESO-SCALE METEOROLOGICAL MODELS IN THE URBAN CONTEXT

Roberto San José and Juan Luis Pérez

TECHNICAL UNIVERSITY OF MADRID

Introduction

This chapter focuses on the meso-scale meteorological models implementation in the urban context (Pielke 1984), with emphasis on the adaptation of meso-scale meteorological models to the case studies that were selected in the framework of the BRIDGE project (Chrysoulakis et al. 2013), as well as on the respective simulations. Meteorological models have been used in studying and assessing the effects of urbanized regions on the meteorological fields. These models, able to resolve meso-scale processes (1–200 km), are considered as important tools in future air pollution assessments, because they allow for sufficiently high spatial and temporal resolution and can trace back the linkages between sources and impacts of long travel distances and times. Additionally they can accommodate a wide range of specific local conditions.

Meso-scale atmospheric models are increasingly employed to improve the understanding of processes that are related to neighbourhood-scale climate and air quality, the Urban Heat Island (UHI), and meso-scale circulations caused by urban land cover. Those processes are strongly influenced by the energy and momentum exchange between the atmosphere and the underlying surface and, hence, the simulation of these processes depends on the accurate characterization of the urban land cover. Meso-scale meteorological models do not have the spatial resolution to directly simulate the fluid dynamics and thermodynamics in and around buildings and other urban structures. They therefore have been adapted to incorporate urban effects, known as "model urbanization".

The planetary boundary layer structure is more complex over urban areas than rural ones, as it consists of canopy and roughness sub-layers not found within typical rural atmospheric surface layers. To capture the grid average effect of detailed urban features in meso-scale atmospheric models, new urban canopy parameterizations into models have been defined and implemented. Urban canopy parameterizations are used to approximate the drag, turbulence production, heating and radiation attenuation induced by sub-grid scale buildings and urban surface covers. The urban canopy layer is composed of diverse individual street canyons, while the urban roughness sub-layer is a non-equilibrium transitions layer, in which vertical momentum, energy, moisture and pollution fluxes from individual urban canyons blend together (Todling & Cohn 1994).

The primary consequence of implementing an urban parameterization in a meso-scale meteorological model is the need to characterize the urban web in detail. Information on urban surface physical properties (e.g. albedo, emissivity) and morphology (e.g. ground elevation, building height and geometry characteristics) is generally needed.

The BRIDGE meso-scale modelling scheme

The modelling approach within BRIDGE integrated different types of models for meso-scale air quality models to urban canopy models. The cascade modelling technique from large to local scale was the main methodology that was applied. This approach allowed estimating the pollutant concentrations and the fluxes associated to different urban development scenarios and strategies.

The Weather Research & Forecasting Model (WRF) meso-scale model was used to simulate meteorological variables through numerical simulations of the atmospheric flows based on the Navier-Stokes equations (Michalakes et al. 2001; Skamarock et al. 2005). WRF uses a high-order accurate discretization scheme for time and space, with new microphysics schemes (Grell et al. 2000). The model is in the public domain and is freely available for community use. It was developed at the National Center for Atmospheric Research (NCAR) which is operated by the University Corporation for Atmospheric Research (UCAR). It is designed to be a flexible, state-of-the-art atmospheric simulation system that is portable and efficient on available parallel computing platforms. It is suitable for use in a broad range of applications across scales ranging from metres to thousands of kilometres. The equation set for WRF is fully compressible (Lorenc 1986), Eulerian and nonhydrostatic with a run-time hydrostatic option (Smolarkiewicz & Margolin 1997). It is conservative for scalar variables. The model uses terrain-following, hydrostatic-pressure vertical coordinate with the top of the model being a constant pressure surface. The horizontal grid is the Arakawa-C grid. The time integration scheme in the model uses the third-order Runge-Kutta scheme, and the spatial discrimination employs 2nd to 6th order schemes. The model supports both idealized and real-data applications with various lateral boundary condition options. The model also supports one-way, two-way and moving nest options. It runs on single-processor, shared- and distributed-memory computers.

WRF simulates the atmospheric flow (meteorology only) in a 3D cube with spatial resolutions on about 1–100 km, with domains between 20–50 km (urban level) and thousands kilometers (continental level) (San José et al. 2008). WRF requires as input the meteorological boundary conditions coming from the outer domain. This information is obtained from the global meteorological model Global Forecast System (GFS), at a 1.0 x 1.0 degree grid, every six hours. They are available on the surface, at 26 mandatory (and other pressure) levels from 1000 hPa to 10 hPa, in the surface boundary layer and at some sigma layers, the tropopause and a few others. The provided parameters include surface pressure, sea level pressure, geopotential height, air temperature, sea surface temperature, soil values, ice cover, relative humidity, u- and v- wind speed components, vertical motion, vorticity and ozone concentration. WRF also requires land cover/use and topographic information, at least at the spatial resolution of its simulations.

Furthermore, the WRF system includes a module called Urban Canopy Model (UCM; Kusaka et al. 2001). Some of the features of the UCM include shadowing from buildings, reflection of short- and long-wave radiation, wind profile in the canopy layer and multi-layer heat transfer equation for roof, wall and road surfaces. UCM describes the partition of surface heat fluxes and the changes in vertical meteorological variables due to the urban canopy layer which is created as a consequence of the urban roughness and the canyon streets (Flanner 2009). UCM has a very thin bucket scheme, i.e. roof and road surfaces are covered with an impermeable layer and the city has an adequate drainage system. Latent heat fluxes from street trees along the road and grass in open spaces are calculated from a single-layer vegetation model. The total latent heat flux from the urban surface is obtained by averaging these fluxes (Kimura 1989). Building height has a significant effect on surface temperature. The larger shadows of taller buildings tend to cool the surface; however, because each surface receives solar and long-wave radiation (Chimklai et al. 2004), taller buildings can also trap the radiative heat. Both the effects of the shadows and the reflected solar and long-wave radiation are included in UCM, which assumes Lambertian surface conditions.

Model implementation

In the framework of BRIDGE, the WRF/UCM model was run for five European cities: Helsinki, Athens, London, Firenze and Gliwice. Two different domains were set: the first domain consisted of 37 x 37 grid cells at 5.4 km x 5.4 km resolution, whereas the second domain consisted of 28 x 28 grid cells at 0.2 km x 0.2 km resolution. The model was run in offline mode, which means that the simulations were performed over a parallel computer platform and then results were integrated in the BRIDGE database.

The WRF/UCM model was setup using the following configuration:

- Cumulus parameterization: Grell 3D ensamble scheme (Grell & Devenyi, 2002). Scheme for higher resolution domains allowing for subsidence in neighbouring columns.
- Planetary Boundary Layer (PBL) scheme and diffusion: Yonsei University (YSU) PBL (Hong & Dudhia 2003). Non-local-K scheme with explicit entrainment layer and parabolic K profile in unstable mixed layer. The YSU PBL scheme is a first order non-local scheme, with a counter gradient term in the eddy-diffusion equation. YSU scheme is dependent upon the bulk Richardson number.
- Explicit moisture scheme: Lin et al. scheme microphysics (Lin et al. 1983). In this scheme six classes of hydrometeors are included: water vapour, cloud water, rain, cloud ice, snow and graupel, suitable for high-resolution simulations.
- Radiation schemes: Rapid Radiative Transfer Model (RRTM) long-wave radiation (Mlawer et al. 1997). It is a spectral-band scheme using the correlated-k method. It uses pre-set tables to accurately represent long-wave processes due to water vapour, ozone, carbon dioxide (CO_2) and trace gases, as well as accounting for cloud optical depth. An accurate scheme using look-up tables for efficiency. Accounts for multiple bands, trace gases and microphysics species. A simple cloud-interactive short-wave radiation scheme, the Dudhia radiation scheme (Dudhia 1989), is available. A simple downward integration, allowing efficiently for clouds and clear-sky absorption and scattering, is also available. It uses look-up tables for clouds.
- Land surface model: Noah/UCM (Chen et al. 2004a). Unified National Centers for Environmental Prediction (NCEP)/NCAR/Air Force Weather Agency (AFWA) scheme with soil temperature and moisture in four layers, fractional snow cover and frozen soil physics. It also includes an option for surface effects for roofs, walls, and streets.

To run the models it is necessary to generate several inputs provided by different sources. Some of them, like emissivity and albedo have been generated uniformly for all BRIDGE case studies, using satellite observations, as described in Chapter 6. Others, such as urban morphology and land cover/use, had to be treated individually in each city, because the original data sources were different. Emissivity and albedo spatio-temporal distributions were available at 1 km x 1 km spatial resolution (Snyder et al. 1998: Schaaf et al. 2002). Figure 7.1 shows an emissivity map for Firenze domain with the value of 0.973 for the built-up area.

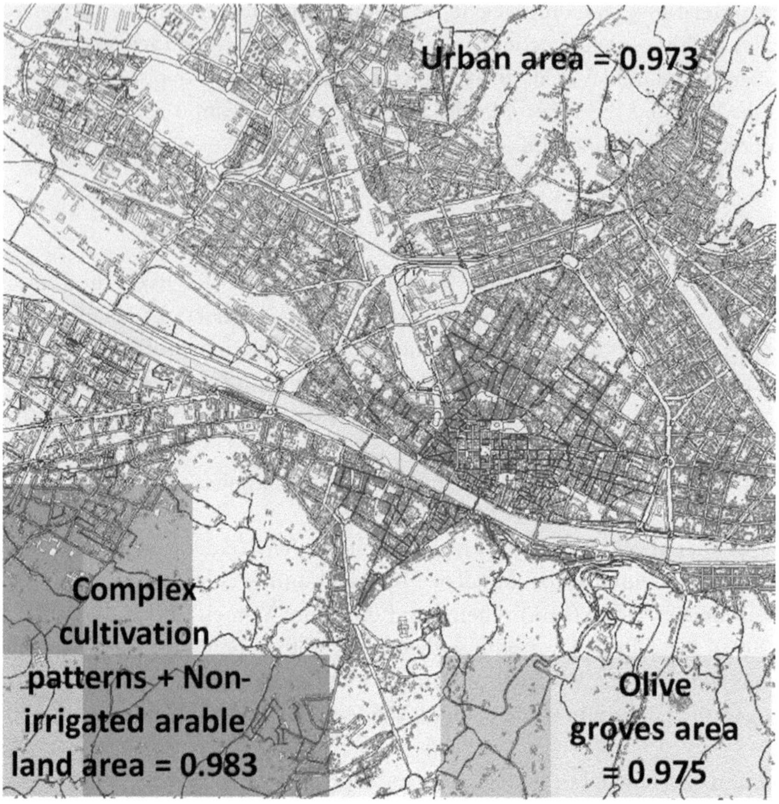

FIGURE 7.1   Emissivity map for Firenze

Albedo changes with the solar zenith angle, thus maps in ten degree intervals (0–90°) were available. To get hourly values (*ALB-h*) the following formula was applied:

$$ALB\_H = (WSA \bullet 0.001) \bullet C + (BSA \bullet 0.001) \bullet (1 - C)$$

where,

BSA (Black Sky Albedo value) is the albedo of the surface if only direct radiation is assumed (available from satellite observations as described in Chapter 6).

WSA (White Sky Albedo value) is the albedo of the surface if only diffused radiation is assumed (available from satellite observations as described in Chapter 6).

C is the fraction of diffused radiation, a function of Solar Zenith Angle (SZA) and Aerosol Optical Thickness (AOT). The parameter C is defined from look-up tables (different for land and sea) if both SZA and AOT are available. The radiation scheme included in the WRF calculates the SZA in hourly basis.

Figure 7.2 shows hourly albedo values for four different AOT conditions, calculated using BSA and WSA maps for Firenze on 16 May 2008.

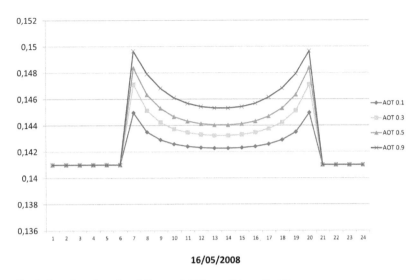

FIGURE 7.2 Albedo hourly values for different AOT conditions for Firenze

Moreover, Corine Land Cover 2000 (CLC2000) data, available from the European Environment Agency (EEA 2012), were used for surface cover parameterization. This data was improved using additional spatial information provided by the city authorities. The CLC2000 uses 44 classes of a three-level CORINE nomenclature. The CLC classification is different from the respective United States Geological Survey (USGS)/UCM used in WRF/UCM. So CLC land cover classes were assigned to USGS-UCM classes, as shown in Table 7.1.

To parameterize surface topography, Advanced Spaceborne Thermal Emission and Reflection Radiometer Global Digital Elevation Model (ASTER GDEM) data at 30 m posting were used. GDEM has been produced using ASTER stereo-imagery (Abrams et al. 2010) and validated for several areas

TABLE 7.1 Correspondence between CLC2000 and USGS/UCM land cover classes

CLC2000 Land Cover Class	USGS/UCM Land Cover Class
Industrial or commercial units	Com/Indust/Transport urban
Discontinuous urban fabric	Low density urban
Continuous urban fabric	High density urban
Sport and leisure facilities	Com/Indust/Transport urban
Road/Rail network	Com/Indust/Transport urban
Green urban area	Grassland
Olive groves	Deciduous broadleaf forest
Irrigated arable land	Dryland cropland/pasture
Complex cultivation patterns	Mixed cropland/pasture
Transitional woodland-shrub	Mixed shrubland/grassland
Forest	Mixed forest
Water courses	Water bodies

worldwide (Chrysoulakis et al. 2011). This data was also improved using additional topographic information and building heights, provided by the city authorities.

Results and validation

Simulations were performed for the full year 2008, for all BRIDGE case studies. In Plates 7 and 8, examples of simulated annual average values of ground heat flux (Wm^{-2}), sensible heat flux (Wm^{-2}), surface runoff (mm) and air temperature (K), at 0.2×0.2 km spatial resolution for Helsinki and Firenze are given.

To ensure that the modelling simulations were scientifically sound, robust and defensible, evaluation procedures were implemented to measure the model's performance. During this evaluation, the modelled estimates were compared against observed values to access the model's accuracy and to provide an indication of its reliability. Emphasis was given on assessing how accurately the model predicted observed values and how accurately the model predicted responses of these values to changes (planning alternatives). The main goal of the evaluation was to determine whether the aggregate modelling scheme (numerical models plus input data sets and observational data for testing) offered sufficiently reliable and accurate results, allowing decision makers to use it in alternative evaluations through the BRIDGE system.

Model validation is needed to confirm that the modelling schemes used perform consistently with their scientific formulation and technical implementation, at a level that is at least as reliable as other current state-of-science methods. Definition of model quality is not a simple task and there is no universal approach. There are no "good" models, or "poor" models, but each model is considered suitable, or not, according to specified objectives. Due to the complexity of the phenomena simulated, there are always uncertainties associated with modelling results. Model evaluation is also complicated by the nature of atmospheric flow and the limited representative observation data.

Despite the wide application of meso-scale meteorological models, no procedures or protocols have been established that assure minimum requirements for the use or performance of such models. Scientific principles that assure model consistency and robustness do exist, but are not organized, or adopted globally according to a well-defined protocol. Several statistical measures, graphical tools and related analytical procedures can be used in the evaluation process. The models' evaluation will utilize numerous graphical displays to facilitate quantitative and qualitative comparisons between predictions and observations. Together with the statistical metrics, the graphical procedures are available to identify obviously flawed model simulations. For example, spatial mean time series plots, time series plots at monitoring locations, outputs maps, output scatter plots stratified by station or by time can be used.

The coupled WRF/UCM model has been applied to major metropolitan regions (e.g. Beijing, Guangzhou/Hong Kong, New York City, Salt Lake City, Taipei, Tokyo and Madrid), and its performance has been evaluated against surface observations, atmospherics soundings, wind profiler data and precipitation data (Chen et al. 2004b; Holt and Pullen 2007; Jiand et al. 2008; Lin et al. 2008; Miao and Chen 2008; Kusaka et al. 2009; Miao et al. 2009a; Wang et al. 2009; Tewari et al. 2010). Results show the coupled WRF/Noah/UCM modelling system is able to reproduce the observed features, reasonably well (Miao & Chen 2008; Miao et al. 2009b). Using the WRF/Noah/UCM scheme significantly improves the simulation of UHI effects and boundary-layer development, when compared to observations obtained from weather stations and Lidars (Lin et al. 2008).

In the framework of BRIDGE, the comparison between observations and simulations gave excellent results for several parameters, such as temperature and heat fluxes. Wind speed comparisons results were found to be poorer than the respective results for temperature and fluxes. Examples of such comparisons are shown in Figures 7.3 and 7.4 for Helsinki and London, respectively.

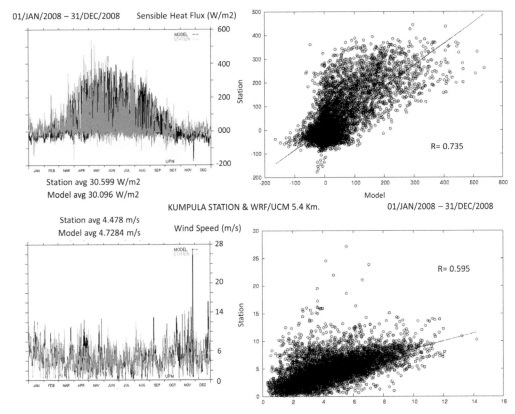

FIGURE 7.3 Full year (2008) time series comparison (left) and lineal regression (right) between WRF/UCM and Kumpula station (Helsinki) measurements: sensible heat flux in Wm^{-2} (top) and wind speed in ms^{-1} (bottom). The comparison concerns the 5.4 km × 5.4 km spatial resolution model domain. The dark grey line is the model.

As shown in Figure 7.4, in London during a few weeks, the simulated wind speed was substantially different from the observed. Several difficulties appeared when trying to run the model for the London area, using the topography provided by the local authorities. Different interpolation procedures had to be employed to overcome this problem. Furthermore, numerical instabilities may be present in these runs, which are probably caused by the extreme height of the buildings and the abrupt changes in topography heights.

Finally, it should be noted that the limitations of computer time were the main cause for not having performed the high spatial resolution simulations (0.2 km) following the nesting rate approach, as required for numerical and stability reasons (0.2 km, 0.6 km, 1.8 km, 5.4 km, 16.2 km, 48.6 km). The computer time required to fulfil this nesting cascade for several years, for five cities and for three different planning alternatives was beyond the supercomputer availability (Barcelona Supercomputer Centre) for the BRIDGE project. An important controversy arose about very high spatial resolution WRF runs (0.2 km, 0.6 km). Numerical issues could be fixed with a substantial increase of vertical layers (higher than 100 vertical layers is suggested by several authors). The 0.2 km spatial resolution seemed to be appropriate, since higher spatial resolution would lead to a completely unreasonable requirement for computer time to perform the simulations, whereas spatial resolution coarser than

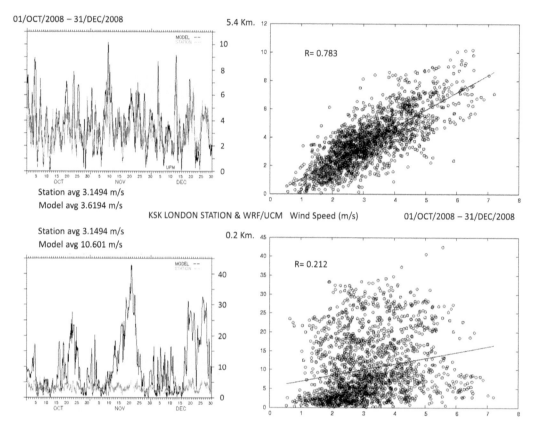

FIGURE 7.4   Time series comparison (left) and lineal regression (right) between WRF/UCM and King's College station (London) measurements: wind speed for the period 1 October to 31 December 2008, at 5.4 km × 5.4 km (top) and 0.2 km × 0.2 km (bottom) resolution domains. The dark grey line is the model.

0.2 km would lead to grid cells where the different local-scale planning alternatives proposed by the city authorities would not have any impact.

Conclusions

How to implement a system of local modelling with high resolution, 0.2 km, has been successfully described. The new tool is a 3D integrated system where the meso-scale model WRF/UCM was included. It is driven by boundary conditions from the global meteorological models. The system has the capability of incorporating *in situ* and remote-sensing data of urban land cover, albedo, emissivity, etc.

In this chapter results from numerical simulations of different European cities are showed and analyzed. Applications of this newly developed modelling system have demonstrated its utility in studying urban meteorology. The WRF/UCM simulations show realistic features; showing great sensitivity to the changes, the system is generally able to capture the influences of urban processes on near-surface meteorological conditions and on the evolution of atmospheric boundary-layer structures in cities. The performance of the modelling system is very good, with average values very close between measurements and modelled and correlation coefficients over 0.7, except during a few weeks over the London urban area.

This methodology allows urban simulations for analysis and study of the urban metabolism (based on energy, water, carbon and pollutant fluxes), which is an essential tool for urban planning and policy makers. The goal was to establish a modelling tool for assessing the impacts of urbanization on urban metabolism by providing accurate meteorological information. The results for analyzing the urban metabolism at high spatial resolution are very promising.

This chapter described a new solution to handle the range of time and length scales which influence urban climate predictions in a computationally tractable way. This involves the use of results from a global meteorological model to force the boundary conditions of a meso-scale model, in which a model of the urban canopy is embedded. These meso-scale predictions may then be used to force the boundary conditions of local micro-scale models, so that good quality local urban climate predictions may be achieved whilst accounting for larger scale climatic tendencies such as the UHI effect.

References

Abrams, M., Bailey, B., Tsu, H. and Hato, M. (2010). The ASTER Global DEM. *Photogrammetric Engineering and Remote Sensing*, 76, 344–348.

Chen, F., Liu, Y., Kusaka, H., Tewari, M., Bao, J.-W., Lo, C.-F. and Lau, K.-H. (2004a). Challenge of Forecasting Urban Weather with NWP Models. 5th MM5 and WRF Users Workshop, 21–25 June, Boulder, Colorado.

Chen, F., Kusaka, H., Tewari, M., Bao, J.-W. and Harakuchi, H. (2004b). Utilizing the coupled WRF/LSM/ urban modeling system with detailed urban classification to simulate the Urban Heat Island phenomena over the Greater Houston area. Preprints, Fifth Symposium on the Urban Environment, Vancouver, BC, Canada, American Meteorological Society, 9–11.

Chimklai, P., Hagishima, A. and Tanimoto, J. (2004). A computer system to support albedo calculation in urban areas. *Building and Environment*, 39, 1213–1221.

Chrysoulakis, N., Abrams, M., Kamarianakis, Y. and Stanisławski, M. (2011). Validation of the ASTER GDEM for the area of Greece. *Photogrammetric Engineering & Remote Sensing*, 77, 157–165.

Chrysoulakis, N., Lopes, M., San José, R., Grimmond, C. S. B., Jones, M. B., Magliulo, V., Klostermann, J. E. M., Synnefa, A., Mitraka, Z., Castro, E., González, A., Vogt, R., Vesala, T., Spano, D., Pigeon, G., Freer-Smith, P., Staszewski, T., Hodges, N., Mills, G. and Cartalis, C. (2013). Sustainable urban metabolism as a link between bio-physical sciences and urban planning: the BRIDGE project. *Landscape and Urban Planning*, 112, 100–117.

Dudhia, J. (1989). Numerical study of convection observed during the winter monsoon experiment using a mesoscale two-dimensional model. *Journal of the Atmospheric Sciences*, 46, 3077–3107.

EEA (2012). The CORINE Land Cover 2000 database. Online. Available HTTP: <http://www.eea.europa. eu/data-and-maps/data/corine-land-cover-2000-clc2000-seamless-vector-database (accessed 7 August 2013)>.

Flanner, M. G. (2009). Integrating anthropogenic heat flux with global climate models. *Geophysical Research Letters*, 36, L02801.

Grell, G. A., Emeis, W. R., Stockwell, T., Schoenemeyer, R., Forkel, J., Michalakes, R., Knoche, R. and Seidl, W. (2000). Application of the multiscale, coupled MM5/chemistry model to the complex terrain of the VOTALP valley campaign. *Atmospheric Environment*, 34, 1435–1453.

Grell, G. A. and Devenyi, D. (2002). A generalized approach to parameterizing convection combining ensemble and data assimilation techniques. *Geophysal Research Letters*, 29, 1693, doi: 10.1029/2002GL0153.

Holt, T. and Pullen, J. (2007). Urban canopy modeling of the New York City metropolitan area: a comparison and validation of single- and multilayer parameterizations. *Monthly Weather Review*, 135, 1906–1930.

Hong, S.-Y. and Dudhia, J. (2003). Testing of a new non-local boundary layer vertical diffusion scheme in numerical weather prediction applications. In: Proceedings of the 16th Conference on Numerical Weather Prediction, Seattle, WA.

Kimura, F. (1989). Heat flux on mixture of different land-use surface: test of a new parameterization scheme. *Journal of the Meteorological Society of Japan*, 67, 401–409.

Kusaka, H., Kondo, H., Kikegawa, Y. and Kimura, F. (2001). A simple-layer urban canopy model for atmospheric models: comparison with multi-layer and slab models. *Boundary-Layer Meteorology*, 101, 329–358.

Kusaka, H., Ohba, M., Suzuki, C., Hayashi, Y., Mizutani, C. (2009). Preliminary analyses of urban heat island phenomenon on a typical winter day in Tsukuba City. *Journal of Heat Island Institute International*, 4, 10–14.

Lin, C. Y., Chen, F., Huang, J. C., Chen, W. C., Liou, Y. A., Chen, W. N. and Liu, S. C. (2008). Urban Heat Island effect and its impact on boundary layer development and land-sea circulation over northern Taiwan. *Atmospheric Environment*, 42, 5635–5649.

Lin, Y. L., Farley, R. D. and Orville, H. D. (1983). Bulk parameterization of the snow field in a cloud model. *Journal of Applied Meteorology*, 22, 1065–1092.

Lorenc, A. C. (1986). Analysis methods for numerical weather prediction. *Quarterly Journal of the Royal Meteorological Society*, 112, 1177–1194.

Miao, S. and Chen, F. (2008). Formation of horizontal convective rolls in urban areas. *Atmospheric Research*, 89, 298–304.

Miao, S., Chen, F., LeMone, M. A., Tewari, M., Li, Q. and Wang, Y. (2009a). An observational and modeling study of characteristics of Urban Heat Island and boundary layer structures in Beijing. *Journal of Applied Meteorology and Climatology*, 48, 484–501.

Miao, S., Chen, F., Li, Q. and Fan, S. (2009b). Impacts of urbanization on a summer heavy rainfall in Beijing, The seventh International Conference on Urban Climate: Proceedings, 29 June – 3 July 2009, Yokohama, Japan, B12–1.

Michalakes, J., Chen, S., Dudhia, J., Hart, L., Klemp, J., Middlecoff, J. and Skamarock, W. (2001). Development of a next generation regional weather research and forecast model. Developments in Teracomputing. In: W. Zwieflhofer and N. Kreitz (Eds). *Proceedings of the Ninth ECMWF Workshop on the Use of High Performance Computing in Meteorology.*

Mlawer, E. J., Taubman, S. J., Brown, P. D., Iacono, M. J. and Clough, S. A. (1997). Radiative transfer for inhomogeneous atmospheres: RRTM, a validated correlated-k model for the longwave. *Journal of Geophysical Research*, 102, 16663–16682.

Pielke, R. A. (1984). *Mesoscale Meteorological Modeling*, Academic Press, New York.

San José, R., Pérez, J. L., Morant, J. L. and González, R. M. (2008). Elevated PM10 and PM2.5 concentrations in Europe: a model experiment with MM5-CMAQ and WRF/CHEM. *WIT Transactions on Ecology and the Environment*, 116, 3–12.

Schaaf, C. B., Gao, G., Strahler, A. H., Lucht, W., Li, X., Tsang, T., Strugnell, N. C., Zhang, X., Jin, Y., Muller, J. P., Lewis, P., Barnsely, M. J., Hobson, P., Disney, M., Roberts, G., Dunderdale, M., Doll, C., D'Entremont, R. P., Hu, B., Liang, S., Privette, J. and Roy, D. P. (2002). First operational BRDF, albedo and nadir reflectance products from MODIS. *Remote Sensing of Environment*, 83, 135–148.

Skamarock, W. C., Klemp, J. B., Dudhia, J., Gill, D., Barker, D., Wang, D. and Powers, J. G. (2005). A description of the advanced research WRF Version 2, NCAR Technical Note NCAR/TN-468+STR.

Smolarkiewicz, P. K. and Margolin, L. G. (1997). On forward-in-time differencing for fluids: an Eulerian/Semi-Lagrangiannon-hydrostatic model for stratified flows. *Atmosphere-Oceans*, 35, 127–152.

Snyder, W. C., Wan, Z., Zhang, Y. and Feng, Y.-Z. (1998). Classification-based emissivity for land surface temperature measurement from space. *International Journal of Remote Sensing*, 19, 2753–2774.

Tewari, M., Kusaka, H., Chen, F., Coirier, W. J., Kim, S., Wyszogrodzki, A. and Warner T. T. (2010). Impact of coupling a microscale computational fluid dynamics model with a mesoscale model on urban scale contaminant transport and dispersion. *Atmospheric Research*, 96, 656–664.

Todling R. and Cohn, S.E. (1994). Suboptimal schemes for atmospheric data assimilation based on the Kalman filter. *Monthly Weather Review*, 122, 2530–2557

Wang, X., Chen, F. Wu, Z., Zhang, M., Tewari, M., Guenther, A. and Wiedinmyer, C. (2009). Impacts of weather conditions modified by urban expansion on surface ozone: comparison between the Pearl River Delta and Yangtze River Delta regions. *Advances in Atmospheric Sciences*, 26, 962–972.

8

URBAN AIR QUALITY MODELS

Carlos Borrego[1], Myriam Lopes[1], Pedro Cascão[1], Jorge Humberto Amorim[1], Helena Martins[1], Richard Tavares[1], Ana Isabel Miranda[1], Matthew James Tallis[2,3] and Peter H. Freer-Smith[3]

[1]UNIVERSITY OF AVEIRO, [2]UNIVERSITY OF PORTSMOUTH AND [3]UNIVERSITY OF SOUTHAMPTON

Introduction

The present chapter concerns the urban air quality modelling conducted under the BRIDGE project (Chrysoulakis et al. 2013). The chapter includes a description of the models and the inputs needed for its application. The models were applied to BRIDGE's case study cities, for a base situation, allowing its evaluation, and also for different urban Planning Alternatives (PA) and Strategic Scenarios devised in the project. The model's results are then presented and the main conclusions deriving from their application are drawn.

Comprehensive Air Quality Model with extensions off-line mesoscale model

Model description

The Comprehensive Air Quality Model with extensions (CAMx) was developed by ENVIRON International Cooperation, from California, United States of America. CAMx (Morris et al. 2004) is an Eulerian photochemical dispersion model that allows the integrated "one-atmosphere" assessment of gaseous and particulate air pollution over many scales ranging from sub-urban to continental. CAMx simulates the emission, dispersion, chemical reaction and removal of pollutants in the troposphere by solving the pollutant continuity equation for each chemical species on a system of nested three-dimensional grids. CAMx carries pollutant concentrations at the centre of each grid cell volume, representing the average concentration over the entire cell. Meteorological fields are supplied to the model to quantify the state of the atmosphere in each grid cell for the purposes of calculating transport and chemistry. CAMx incorporates two-way grid nesting, which means that pollutant concentration information propagates into and out of all grid nests during model integration. The nested grid capability of CAMx allows cost-effective application to large regions in which regional transport occurs, yet at the same time providing fine resolution to address small-scale impacts in selected areas (ENVIRON 2010). The CAMx gas-phase chemical mechanisms include the 2005 version of Carbon Bond (CB05, Yarwood et al. 2005), SAPRC99 (Carter 2000) and several versions of Carbon Bond IV (CB4, Gery et al. 1989).

Architecture and domains

CAMx model was applied to three of the five BRIDGE study cases (see Chapter 3 for details) – Helsinki, Gliwice and London – and only to the base situation, since the impacts of the different

PA in terms of air quality will be analysed by the Urban Air Quality (URBAIR) model in the next section. The air quality simulations were performed for two domains for each city, corresponding to the Weather Research and Forecasting (WRF) meteorological simulation domains. The first domain (with 5.4 km × 5.4 km horizontal resolution) was used to drive the second domain's simulations (0.2 km × 0.2 km horizontal resolution). Both domains were simulated with 17 vertical levels.

Due its complexity and computer resources needs, CAMx model will not integrate the BRIDGE Decision Support System (DSS); the model was run separately and the output data are stored in the DSS system database. The simulations were performed on a 25-node Beowulf Linux Cluster belonging to the University of Aveiro.

Input data preprocessing

CAMx requires input files that configure each simulation, define the chemical mechanism, and describe the photochemical conditions, surface characteristics, initial/boundary conditions, emission rates and various meteorological fields over the entire modelling domain. Preparing this information requires several preprocessing steps to translate "raw" emissions, meteorological, air quality and other data into the final input files for CAMx. Some changes have been performed in order to implement WRFCAMx system.

The WRFCAMx preprocessor generates CAMx meteorological input files from the WRF output files, including land use, altitude/pressure, wind, temperature, moisture, clouds/rain and vertical diffusivity. Land use information was also provided by the WRF model through the WRFCAMx preprocessor. However, the program considered only the 24 United States Geological Survey (USGS) land use categories originally present in WRF model; WRFCAMx was altered in order to consider the additional land use classes: low density urban, high density urban and commercial/industrial and transportation.

Initial concentrations and hourly boundary conditions were created from output concentration files from the Laboratoire de Météorologie Dynamique – Zoom-Interaction with Chemistry and Aerosols (LMDz-INCA) chemistry-climate global circulation model (Hauglustaine et al. 2004) for gaseous species, and from the global model Goddard Chemistry Aerosol Radiation and Transport (GOCART; Ginoux et al. 2001) for aerosols.

CAMx air quality simulations for Helsinki (both domains), Gliwice (both domains) and London's domain 1, with 5.4 km × 5.4 km resolution, were performed using the TNO (The Netherlands Organisation for Applied Scientific Research) European emission data set prepared for the European Union (EU) Framework Programme (FP) project Global and regional Earth-system (Atmosphere) Monitoring using Satellite and in-situ data (GEMS), for the year 2003, with 1/8 × 1/16 degree resolution (Visschedijk et al. 2007). The inventory includes emissions for sulfur dioxide (SO_2), nitrogen oxides (NO_x), ammonia (NH_3), carbon monoxide (CO), non-methane volatile organic compounds (NMVOC), methane (CH_4), particulate matter with aerodynamic diameter smaller than 10 μm (PM10) and 2.5 μm (PM2.5), given at the Standardized Nomenclature for Air Pollutants (SNAP) 97 1st level that consists of 11 source categories, commonly adopted in air quality modelling (Visschedijk et al. 2007). In addition, the inventory also includes emission data for international shipping. However, for the high-resolution air quality simulations in BRIDGE, the above mentioned inventory had to be further spatially disaggregated. The spatial disaggregation of emissions was done according to Corine Land Cover 2006 data with 100 m resolution (www.eea.europa.eu/data-and-maps/data/corine-land-cover-2006-clc2006-100-m-version-12-2009), following the methodology developed by Maes et al. (2009).

The London Atmospheric Emissions Inventory (LAEI) was used for the air quality simulation of London's domain 2. Although it was not possible to use it for the larger domain, as the LAEI does not cover the entire domain, it was decided that it would be used for the smaller one due to its high spatial resolution. The LAEI is a database with information on emissions from all sources of air pollutants in the Greater London area The LAEI includes detailed emission information on several air pollutants including: NO_x, SO_2, CO, NMVOC, PM10 and PM2.5. Traffic emissions are provided on a road link basis for the major roads; other sources, including domestic and commercial emissions, minor roads, airport, rail and shipping, are given on a 1 km × 1 km grid square basis. To further spatially disaggregate the LAEI to the 200 m x 200 m resolution the above mentioned methodology was once more applied.

Output data

CAMx post-processors allow the extraction of time series simulated concentrations for predefined locations and bi-dimensional concentration fields for a given pollutant. CAMx simulations supplied three data sets for the DSS database, one for each city, that were made available for on-line models and that were also be used to calculate the indicators. The data sets include hourly gridded concentrations for ozone (O_3), nitrogen dioxide (NO_2), PM10 and PM2.5 for each city case's smallest domain with 200 m × 200 m spatial resolution, for the entire year of 2008, in a total of 79056 files (3 cities × 3 pollutants × 366 days × 24 hours), corresponding to around 12 GB.

Results

This section includes a brief analysis of the models' results and the values obtained for the indicators established for BRIDGE project regarding air quality. Plate 9 presents the annual average concentrations for NO_2, PM10 and PM2.5 for Gliwice's domain 2. Both NO_2 and PM10 present annual concentrations considerably below the legislated values (2008/50/EC): 40 µg.m^{-3} for NO_2 and 48 µg.m^{-3} (40 + 20 per cent tolerance) for PM10. As for Gliwice, in Helsinki all the pollutants present annual concentrations considerably below the legislated values. In London, simulated PM10 concentrations were considerably higher than those for Helsinki and Gliwice, with the annual limit values exceeded in certain locations of the simulation domain.

In the BRIDGE project a set of indicators of environmental, socio-economic and sustainability impacts were defined. Table 8.1 presents the air quality indicator values obtained with CAMx for its three study cases.

TABLE 8.1 Air quality indicators produced by CAMx for Helsinki, Gliwice and London

Indicator name	Indicator value			
	Threshold	Helsinki	Gliwice	London
Annual (NO_2)	50 µg.m^{-3}	25 µg.m^{-3}	20 µg.m^{-3}	–
Annual (PM10)	48 µg.m^{-3}	11 µg.m^{-3}	28 µg.m^{-3}	14 µg.m^{-3}
Annual (PM2.5)	no limit defined	10 µg.m^{-3}	24 µg.m^{-3}	9 µg.m^{-3}
Maximum 8-h (O_3)	120 µg.m^{-3}	107 µg.m^{-3}	113 µg.m^{-3}	149 µg.m^{-3}
No. of NO_2 exceedances	200 µg.m^{-3} not to be exceeded more than 18 times a year	0	0	–
No. of PM10 exceedances	50 µg.m^{-3} not to be exceeded more than 35 times a year	0	13	7
No. of O_3 exceedances	120 µg.m^{-3} for 8h averages not to be exceeded more than 25times a year	0	0	3

Model performance evaluation

In this section CAMx results obtained for the Helsinki and Gliwice study cases are compared against observations coming from local air quality monitoring stations. A group of statistical indicators is also presented to allow the evaluation of the model's performance.

The following statistical indicators have been calculated (Thunis et al. 2010):

- The Correlation Coefficient (R): ranging from −1 to +1, indicates the strength of a linear relationship between two data sets (in this case, observed and modelled). A value of R near to zero indicates the absence of linear correlation between the variables.

$$R = \sum_{i=1}^{N}\left(M_i - \overline{M}\right)\cdot\left(O_i - \overline{O}\right) / \sqrt{\sum_{i=1}^{N}\left(M_i - \overline{M}\right)^2}\cdot\sqrt{\sum_{i=1}^{N}\left(O_i - \overline{O}\right)^2}$$

- The Index of Agreement (IOA): determines the extent to which magnitudes of observed mean values are related to the predicted deviations and allows for sensitivity toward differences in observed and modelled values; the perfect value of IOA is 1.

$$IOA = 1 - N\cdot RMSE^2 / \sum_{i=1}^{N}\left(\left|M_i - \overline{O}\right| + \left|O_i - \overline{O}\right|\right)^2$$

- The Mean Fractional Bias (MFB): is a useful indicator because it has the advantage of equally weighting positive and negative bias estimates and does not assume that observations are the truth (i.e. the denominator is the sum of observed and predicted); it ranges from −200 per cent to +200 per cent.

$$MFB = \frac{1}{N}\sum_{i=1}^{N}\frac{M_i - O_i}{\left(M_i + O_i\right)/2}$$

- The Mean Fractional Error (MFE): gives equal weight to under- and over-prediction and is not sensitive to a threshold in observed values; it ranges from 0 per cent to +200 per cent.

$$MFE = \frac{1}{N}\sum_{i=1}^{N}\frac{\left|M_i - O_i\right|}{\left(M_i + O_i\right)/2}$$

- Factor of modelled values within a factor of two of observations (FAC2).

$$FAC2 = \frac{1}{N}\sum n_i \text{ with } n_i = \begin{cases} 1 \text{ for } 0.5 \le \left|M_i / O_i\right| \le 2 \\ 0 \text{ else} \end{cases}$$

- The Relative Directive Error (RDE): defined in relation to the Air Quality Directive (AQD) 2008 in order to give a mathematical expression of the "model uncertainty" term in the AQD (where OLV is the closest observed concentration to the limit value concentration (LV) and MLV is the correspondingly ranked modelled concentrations.

$$RDE = \frac{\left|O_{LV} - M_{LV}\right|}{LV}$$

TABLE 8.2 Model quality objectives for O_3, PM10 and NO_2

	MFB	MFE	FAC2	R	IOA	RDE	RPE
O_3	<0.3	<0.45	>0.5	>0.65	>0.65	<0.5	<0.5
PM10	<0.6	<0.75	>0.5	>0.4	>0.6	<0.5	<0.5
NO_2	<0.4	<0.5	>0.5	>0.25	>0.5	<0.5	<0.5

Source: Thunis et al. 2010.

- The Relative Percentile Error (RPE): an alternative model error measure for the purposes of the AQD, defined as the concentration difference at the percentile p corresponding to the allowed number of exceedances of the limit value normalized by the observation.

$$RPE = \frac{\left| O_p - M_p \right|}{O_p}$$

Thunis et al. (2010) propose the model quality objectives presented in Table 8.2.

Gliwice

CAMx results for Gliwice are compared against observations for 2008 obtained from AIRBASE (http://acm.eionet.europa.eu/databases/airbase/) for a local air quality monitoring station – Mewy (code PL0238A) – classified as urban background, which measures PM10 and NO_2, and is located at 18.7° East and 50.3° North. The statistical parameters obtained with CAMx for PM10 and NO_2 are presented in Table 8.3 (values in bold meaning that they do not accomplish the model quality objectives proposed by Thunis et al. (2010) presented in the previous table). It is possible to conclude that for both pollutants the majority of the statistical parameters obtained the quality objective is accomplished, with a slightly better performance obtained for PM10.

Helsinki

CAMx results for Helsinki are compared against observations for 2008 obtained from AIRBASE (http://acm.eionet.europa.eu/databases/airbase/) for a local air quality monitoring station – Kalio (code FI00425) – classified as urban background, which monitors PM10, O_3 and NO_2. Figure 8.1 presents the scatter plots for PM10, O_3 and NO_2, modelled and observed.

The statistical parameters obtained with CAMx for PM10, O_3 and NO_2 are presented in Table 8.4). All pollutants comply with the statistical parameters quality objectives, except for R and the IOA. O_3 is the pollutant for which the better performance is obtained.

TABLE 8.3 Statistical parameters obtained with CAMx for Gliwice – Mewy (values in bold do not accomplish the quality objective)

	MFB	MFE	FAC2	R	IOA	RDE	RPE
PM10	0.26	0.48	0.71	**0.35**	**0.49**	0.25	0.5
NO_2	**0.5**	**0.57**	0.5	0.3	**0.47**	0.30	0.46

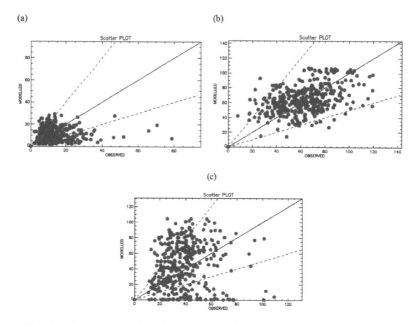

FIGURE 8.1 (a) PM10 daily mean, (b) O_3 daily 8-hour maximum and (c) NO_2 daily maximum scatter plot for Helsinki – Kalio

TABLE 8.4 Statistical parameters obtained with CAMx for Helsinki – Kalio (values in bold do not accomplish the quality objective)

	MFB	MFE	FAC2	R	IOA	RDE	RPE
PM10	0.22	0.60	0.61	**0.2**	**0.4**	0.49	0.50
O_3	0.05	0.29	0.92	**0.5**	0.65	0.1	0.1
NO_2	0.06	**0.65**	0.59	**0.15**	**0.42**	0.02	0.02

URBAIR on-line urban scale model

Within the scope of BRIDGE, and specifically with the purpose of developing a DSS for sustainable urban planning, it proved important to develop and implement an on-line air quality model capable of providing air pollutant levels at the urban scale, with higher spatial resolution than the one obtained with mesoscale models, but with much lower computational effort than the one resulting from the application of a Computational Fluid Dynamics (CFD) model. This effort culminated with the development of the URBAIR model (Borrego et al. 2011). This on-line second generation Gaussian model integrates a number of functionalities, namely the estimation of road traffic emissions, and the simulation of the resulting air quality patterns for a given spatial domain and time period (usually one year, in compliance with the Directive) for the following air pollutants: PM10, NO_2, SO_2 and CO. Because of the capability to simulate the effect of buildings on air pollutant dispersion, URBAIR offers the possibility to assess the impact of urban planning strategies and traffic management scenarios on air quality.

Model structure

URBAIR model integrates the preprocessing of urban morphology, meteorological information and air pollutant emission data in a single tool specifically developed to run online in a DSS build under

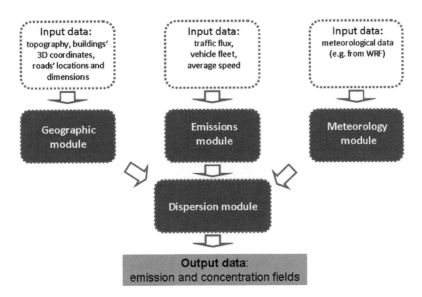

FIGURE 8.2 URBAIR model STRUCTURE

a Geographical Information System (GIS) platform. The URBAIR structure is organized into four modules as schematically shown in Figure 8.2. It also includes a series of other preprocessors that prepare the input data, namely in terms of format compatibility.

The estimation of road traffic emissions is performed with the Transport Emission Model for Line Sources (TREM) (Borrego et al. 2003), which has been integrated into URBAIR under the scope of BRIDGE. The transport and dispersion of the emitted air pollutants (gaseous and particles) is modelled by applying an improved version of the second generation Gaussian model POLARIS (Borrego et al. 1997). This model is significantly different from traditional Gaussian dispersion models, because it considers the presence of buildings during the dispersion simulation and its dispersion parameters have a continuous variation with the atmospheric stability. URBAIR is suitable to be used for distances up to about 10 km from the domain centre, and is based in a steady state atmospheric dispersion model. The modelling approach implemented follows the Monin–Obukhov similarity theory approach (Tavares et al. 2011). For this purpose, a preprocessor calculates the meteorological parameters needed by the dispersion model, namely the atmospheric turbulence characteristics, mixing heights, friction velocity, Monin–Obukov length and surface heat flux. Furthermore, URBAIR requires meteorological information that is provided by the mesoscale meteorological model WRF. Alternatively, surface measurements and upper air soundings databases can also be used.

Topography and build-up structures' characteristics have a significant influence on the dispersion of atmospheric pollutants, in particular in urban areas. In this sense, URBAIR also requires the characterization of the spatial variation of terrain surface elevation, buildings' 3D coordinates and emission sources' locations and dimensions, which are usually provided by GIS maps. The geographic module relies on a cartesian coordinate system in which regular and discrete gridded data can be used to characterize and spatially distribute terrain, receptors and sources. Topography is specified in the form of terrain heights at receptor locations. The influence of buildings on air pollutant dispersion depends on the orientation of the building with respect to the source, the wind direction and the shape of the building. Direction-specific downwash parameters, in the form of projected building height and width dimensions, are estimated using the EPA's Building Profile Input Program - Plume Rise Model Enhancements (BPIP-PRIME) modelling approach (Schulman et al. 2000).

Input/output data

As previously mentioned the major categories of input data needed by the integrated air quality system URBAIR are the meteorological conditions, the geographic/geometric characteristics and road traffic fluxes. The methodology adopted to preprocess the input data for URBAIR starts with the definition of the study area based on the information relating the proposed PA and using GIS-based maps (see Chapter 3 for details). Due to the high number of buildings within the study areas (which are in the order of some thousands), buildings were grouped in blocks that are delimited by adjacent roads, resulting in a great reduction of computational effort.

Traffic is considered as the main pollutant source in the study areas. Emissions are calculated by the preprocessor TREM using traffic counts and average speeds provided by each city. Meteorological input data, including vertical profiles, were obtained from the WRF mesoscale model simulations over the different case studies domains. The meteorological data from WRF required to run URBAIR is divided in two categories: surface meteorological parameters and vertical profiles. These parameters include: location latitude and longitude, date and time, sensible heat flux, surface roughness length, albedo, wind speed and direction, temperature, precipitation, relative humidity and air pressure. Because the meteorological output data from WRF was not in a format that URBAIR could use, a preprocessor was built to extract and preprocess the input data according to URBAIR requirements. Finally, the output data produced by URBAIR includes the meteorological parameters and pollutant concentration at user-specified receptor points or spatially distributed over a regular grid.

Implementation of PAs

A summary of the simulation domain characteristics for the set of case studies and correspondent scenarios is presented in Table 8.5.

In the case of Firenze and London, the PA are only related to meteorology changes and do not involve modifications in the urban structure. For this reason, its implementation in URBAIR only requires the modification of the meteorological input files.

Air quality results for baseline and PAs

As an example for a specific summer day in the Athens study case, Plate 10 presents the simulation results for PM10 in an intercomparison between the baseline and the different PAs. Please note that PA1 is only related to modifications in meteorology (see Chapter 3 for details).

Analysing the results presented in Plate 10, it is clear that PA3 (conversion to green area) is the one that presents the most significant reduction of PM10 levels in the intervention area as a consequence of the decreased traffic flow. According to the URBAIR simulations, PA1 (meteorological changes) and PA2 (urban fabric) have little impact over the dispersion pattern and magnitude of the concentrations

TABLE 8.5 Simulation domain characteristics, for baseline and PAs 1, 2 and 3

Study case	Simulation area (m²)	Spatial grid resolution (m²)	Number of rearranged blocks			
			Baseline	PA1	PA2	PA3
Helsinki	4000 × 4000	100 × 100	234	251	254	263
Athens			151	–	165	118
Gliwice			92	93	93	95

attained when compared with the baseline. In the case of Helsinki, the comparison of results gave rise to the conclusion that despite the changes in the number of roads and respective traffic fluxes, and also in the number and location of buildings, the different alternatives do not induce significant modifications over the dispersion patterns. However, and according to the simulations, PA2 and PA3 (see Chapter 3 for details) have a higher influence over the PM10 levels in the intervention area and, particularly in PA3, in an area located to the north of the new buildings and roads. In Gliwice, no major differences in PM10 concentration were obtained as a result of the Sports Centre construction (PA1), mostly because of its seasonal use, when compared with the baseline situation. On the contrary, the Centre for New Technologies would lead to an increase on PM10 levels in PA2, and particularly in PA3 in conjunction with the Sports Centre, as a consequence of the increase in traffic flows in the nearby roads (Borrego et al. 2011). Comparing the air quality results for Firenze and since the modifications to PA1, PA2 and PA3 are only related to modifications in the meteorology (see Chapter 3 for details), there aren't significant modifications in PA1 and PA3, and in PA2 the PM10 concentration field has lower values when compared with the other cases. In London, a PM10 concentration hotspot is clearly visible in all the situations. No major differences in the concentrations and dispersion patterns were observed with the different PAs.

Model performance evaluation

The air quality results obtained with URBAIR for the baseline situation were analysed and evaluated using observed data when possible. As an example, Figure 8.3 presents a time series of simulated and measured PM10 concentrations for the year 2008 in Athens. Observed air quality levels were acquired at the Aristotelous Air Quality monitoring Station (AQS). The simulated values were extracted from a specific cell of the domain which corresponds to the location of the referred AQS. The background concentrations monitored in an AQS located outside the study domain were added to the URBAIR simulated concentrations.

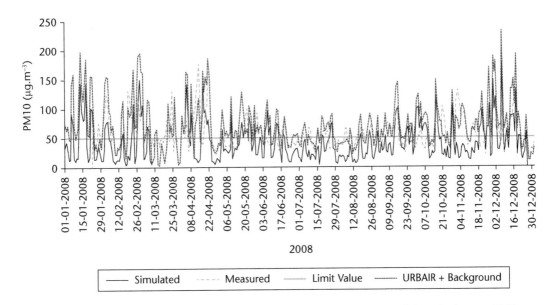

FIGURE 8.3 Comparison of measured versus simulated PM10 concentration in Athens in the year 2008 (at coordinates x, y = 2800 m, 2000 m)

TABLE 8.6 URBAIR performance statistics obtained for PM10 and NO$_2$

Study case	Pollutant	R	IOA
Athens	PM10	0.6	0.6
	NO$_2$	0.4	0.5
Firenze	PM10	0.8	0.7
London	PM10	0.4	0.6

Note: The URBAIR outputs account for the contribution of background air quality levels.

In general, simulated values reasonably follow the trend of measured concentrations. Also, both measured and simulated PM10 concentrations exhibit significantly high values when compared (with the needed caution given the different time averages) with the annual average limit value of 50 µg m^{-3} given by legislation.

The quality indicators IOA and R have been used to evaluate the URBAIR performance over the study region. In Table 8.6 the statistical parameters obtained for the pollutants PM10 and NO$_2$ are presented for the study cases where observed data is available.

From the statistical analysis of modelling performance shown in Table 8.6 it can be concluded that all the results accomplish the quality objectives proposed by Thunis et al. (2010). In all the cases, adding the contribution of urban background concentration to the values simulated by URBAIR has improved the results' accuracy, which is in agreement with previous studies (e.g. Tchepel et al. 2010). For the Helsinki case study, modelling quality indicators were not calculated because of the inexistence of an AQS in the domain. For Gliwice, the AQS is located too close to the boundaries to allow an adequate comparison between simulated and measured concentrations. For this reason it was decided not to use the observed data for Quality Assurance/Quality Control (QA/QC) purposes.

Pollution deposition flux model

Model description

Urban greenspace is a receptor for airborne particulate pollution. Here we developed a modelling approach to estimate the annual PM10 removal by the urban tree canopy of London now and for future planting scenarios. The flux deposition model is developed from Nowak & Crane (1998) using species specific deposition parameters as described in Tiwary et al. (2009) and adapted to accept high resolution city-wide maps of PM10 dispersal, for example, the LAEI maps described earlier. The model is further described in Tallis et al. (2011). To estimate the PM10 removal using this flux deposition model the following are needed: (i) urban canopy area (including species and leaf area index; if not available then a simple classification as broadleaved deciduous or coniferous evergreen), (ii) PM10 concentration local to the canopy and (iii) meteorological data.

Model application to London

The urban tree canopy for London was classified (street tree, urban woodland or garden tree) and quantified using maps generated from Earth Observation data (Fuller et al. 2001) and results from ground surveys, aerial photography (Britt & Johnston 2008) and street tree inventories (GLA 2007). Monthly meteorological data were extracted from 5 km gridded data sets mapped to the Greater London

TABLE 8.7 Flux model evaluation

Site	PM10 concentration (µg m⁻³)	Measured PM10 deposition (g m⁻²) (± 1 SD)	Modelled PM10 deposition (g m⁻²)
Low	31	0.032 (0.01)	0.061
Medium	39	0.059 (0.03)	0.076
High	47	0.060 (0.02)	0.092

Authority (GLA) (Perry & Hollis 2005). Deposition for the whole GLA was calculated from the sum of model runs for each canopy type and adapted to account for future planting schemes, climates and PM10 concentration (Tallis et al. 2011).

Model results and evaluation

The mean modelled rates of annual PM10 deposition to the whole urban tree canopy of London varied from 27.3 kg ha⁻¹ y⁻¹ to 58.5 kg ha⁻¹ y⁻¹ (depending on modelling assumptions) and fall within the range (11 to 80 kg ha⁻¹) for annual PM10 removal by trees and shrubs modelled for 55 USA cities (Nowak et al. 2006).

Within the modelled area, sites with measured hourly PM10 concentration representing a range in concentrations were selected (http://www.londonair.org.uk). At these sites leaves from *Acer platanoides* and *Platanus* × *acerifolia* were sampled. The mass of non-soluble particles deposition on the leaves for a period of 56 days was measured using the protocol described in Beckett et al. (2000), and compared with the modelled values for the same time period parametrised for deposition to *Acer pseudoplatanus* using values from Freer-Smith et al. (2004). Leaves of these deciduous trees were clean at the start (1 April 2001) as this was the assumed bud-flush date. These initial results support the model well for these species and pollution environments and are reported in Table 8.7.

References

Beckett, K.P., Freer-Smith, P.H. & Taylor, G. (2000). Effective tree species for local air-quality management. *Journal of Arboriculture*, 26, 12–19.

Borrego, C., Martins, J.M, Lemos, S. & Guerreiro, C. (1997). Second generation Gaussian dispersion model: the POLARIS model. *International Journal of Environment and Pollution*, 8, No.3/4/5/6, 789–795.

Borrego, C., Tchepel, O., Costa, A., Amorim, J. & Miranda, A. (2003). Emission and dispersion modelling of Lisbon air quality at local scale. *Atmospheric Environment*. 37, 5197–5205.

Borrego C., Cascão P., Lopes M., Amorim J.H., Tavares R., Rodrigues V., Martins J., Miranda A.I. & Chrysoulakis N. (2011). Impact of urban planning alternatives on air quality: URBAIR model application. C.A. Brebbia and J.W.S. Longhurst (Eds). In: *Air Pollution XIX*. WIT Transactions on Ecology and the Environment 147. 12.

Britt, C. & Johnston, M. (2008). Trees in towns II: a new survey of urban trees in England and their condition and management. *Research for Amenity Trees*; No. 9. Department for Communities and Local Government.

Carter, W. P. L. (2000). Implementation of the SAPRC-99 Chemical Mechanism into the Models-3 Framework. Report to the United States Environmental Protection Agency, January 29. Available at http://www.cert.ucr.edu/~carter/pubs/s99mod3.pdf

Chrysoulakis, N., Lopes, M., San José, R., Grimmond, C.S.B., Jones, M.B., Magliulo, V., Klostermann, J.E.M., Synnefa, A., Mitraka, Z., Castro, E., González, A., Vogt, R., Vesala, T., Spano, D., Pigeon, G., Freer-Smith, P., Staszewski, T., Hodges, N., Mills, G., & Cartalis, C. (2013). Sustainable urban metabolism as a link between bio-physical sciences and urban planning: the BRIDGE project. *Landscape and Urban Planning*, 112, 100–117.

ENVIRON (2010). *CAMx v5.4.1 User's Guide*. ENVIRON International Corporation, Novato, California, EUA. June.

Freer-Smith, P.H., El-Khatib, A.A. & Taylor, G. (2004). Capture of particulate pollution by trees: a comparison of species typical of semi-arid areas (*Ficus nitida* and *Eucalyptus globulus*) with European and North American species. *Water, Air and Soil Pollution*, 155, 173–187.

Fuller, R., Smith, G.M., Sanderson, J.M., Hill, R.A. & Thompson, A. G. (2001). The UK Land Cover Map 2000: construction of a parcel-based vector map from satellite images. *Cartographic Journal*, 39, 15–25.

Gery, M.W., Whitten, G.Z., Killus, J.P. & Dodge, M.C. (1989). A photochemical kinetics mechanism for urban and regional scale computer modeling. *Journal of Geophysical Research*, 94, 925–956.

Ginoux, P., Chin, M., Tegen, I., Prospero, J.M., Holben, B., Dubovik, O. & Lin., S.J. (2001). Sources and distributions of dust aerosols simulated with the GOCART model. *Journal of Geophysical Research*, 106, 20255–20274.

GLA (2007). Chainsaw massacre: a review of London's street trees. Greater London Authority.

Hauglustaine, D.A., Hourdin, F., Walters, S., Jourdain, L., Filiberti, M., Larmarque, J. & Holland, E. (2004). Interactive chemistry in the Laboratoire de Météorologie Dynamique general circulation model: description and background tropospheric chemistry evaluation. *Journal of Geophysical Research*, 109, D04314, doi:10.1029/2003JD003957.

Maes, J., Vliegen, J., Vel, K.V., Janssen, S., Deutsch, F., Ridder, K. & Mensink, C. (2009). Spatial surrogates for the disaggregation of CORINAIR emission inventories. *Atmospheric Environment*, 43, 1246–1254.

Morris, R.E., Yarwood, G., Emery, C. & Koo, B. (2004). *Development and Application of the CAMx Regional One-Atmosphere Model to Treat Ozone, Particulate Matter, Visibility, Air Toxics and Mercury*. Presented at 97th Annual Conference and Exhibition of the A&WMA, June 2004, Indianapolis.

Nowak, D.J. & Crane, D.E. (1998). The Urban Forest Effects (UFORE) model: quantifying urban forest structure and functions. *Integrated Tools proceedings*.

Nowak, D.J., Crane, D.E. & Stevens, J. C. (2006). Air pollution removal by urban trees and shrubs in the United States. *Urban Forestry & Urban Greening*, 4, 115–123.

Perry M. & Hollis D. (2005). The generation of monthly gridded datasets for a range of climatic variables over the UK. *Interernational Journal of Climatology*, 25, 1041–1054.

Schulman, L.L., Strimaitis, D.G. & Scire, J.S. (2000). Development and evaluation of the PRIME plume rise and building downwash model. *Journal Air Waste Management Association*, 50, 378–390.

Tallis M.J., Freer-Smith P.H., Sinnett D. & Taylor G. (2011). Estimating the removal of atmospheric particulate pollution by the urban tree canopy of London, under current and future environments. *Landscape and Urban Planning*, 103, 129–138.

Tavares, R., Miranda, A.I. & Borrego, C. (2011). Modelling and assessing risks from accidental release of hazardous gases. *Proceedings of the 2nd International Conference on Air Pollution and Control*, 19–23 September, Antalya, Turkey.

Tchepel, O., Costa, A.M., Martins, H., Ferreira, J., Monteiro, A., Miranda, A.I. & Borrego, C. (2010). Determination of background concentrations for air quality models using spectral analysis and filtering of monitoring data. *Atmospheric Environment*, 44, 106–114.

Thunis, P., Georgieva, E. & Galmarini, S. (2010). *A procedure for air quality models benchmarking – version 1* (draft). Joint Research Centre, Ispra.

Tiwary A., Sinnett, D., Peachy, C., Chalabi, Z., Vurdoulakis, S., Fletcher, T., Leonardi, G., Grundy, C. & Hutchings, T.R. (2009). An integrated tool to assess the role of new planting in PM_{10} capture and the human health benefits: a case study in London. *Environmental Pollution*, 157, 2645–2653.

Visschedijk, A.J.H., Zandveld, P. & Denier van der Gon, H.A.C. (2007). A high resolution gridded European emission database for the EU integrated project GEMS. *TNO report 2007-A-R0233/B*.

Yarwood, G., Whitten, G.Z. & Rao, S. (2005). *Updates to the Carbon Bond 4 Photochemical Mechanism*. Prepared for the Lake Michigan Air Directors Consortium, Des Plains, Illinois.

9

URBAN ENERGY BUDGET MODELS

C.S.B. Grimmond[1], Leena Järvi[3], Fredrik Lindberg[4,2], Serena Marras[5,9], Matthias Falk[6,9], Thomas Loridan[7,2], Gregoire Pigeon[8], David R. Pyles[6,9] and Donatella Spano[5,9]

[1]UNIVERSITY OF READING, [2]KING'S COLLEGE LONDON, [3]UNIVERSITY OF HELSINKI, [4]UNIVERSITY OF GÖTEBORG, [5]UNIVERSITY OF SASSARI, [6]UNIVERSITY OF CALIFORNIA, [7]RMS, [8]METEO FRANCE & CNRS, [9]EURO-MEDITERRANEAN CENTER ON CLIMATE CHANGE

Introduction

As part of the BRIDGE project (Chrysoulakis et al. 2013), the urban energy balance (UEB) fluxes (Chapter 4) were measured (Chapter 5) and modelled, independently (offline) and nested in meso-scale models (Chapter 7). In this chapter we consider the models used and their performance. Knowing how well models perform is critical if they are to be used to evaluate the influence of proposed changes in an urban area (Chapter 17), such as adding trees, and their influence on the surface energy, water (Chapter 10), carbon exchanges (Chapter 11) and air quality (Chapter 8).

A large number of UEB models now exist that are able to simulate urban surface–atmosphere exchanges (for a recent review see Grimmond et al. 2009). There is no single "perfect" urban land surface model (Grimmond et al. 2010, 2011), hence a powerful methodology is to use a range of models (an ensemble approach) to consider probable impacts of changes to the urban surface on atmospheric conditions.

Here we consider five different models (see "Models used in BRIDGE"), run by different research groups using slightly different set-ups for each of the case study cities (see "Application and results"). First, we describe the models and then assess their performance.

Models used in BRIDGE

The five urban land surface models used to investigate urban energy balance fluxes in the five BRIDGE cities studied (Firenze, Helsinki, London, Gliwice, Athens) were:

1. Advanced Canopy-Atmosphere-Soil Algorithm (ACASA);
2. Noah land surface model and the Single Layer Urban Canopy Model (Noah/SLUCM);
3. Surface Urban Energy and Water Balance Scheme (SUEWS);
4. Local scale Urban Parameterization Scheme (LUMPS);
5. Town Energy Balance (TEB)

In addition, for anthropogenic heat flux, the Large scale Urban Consumption of energY model (LUCY) was used.

All of the models are fully described in papers in the refereed literature (referred to in the following subsections); hence most details are not repeated here. Some of the models have been extensively evaluated (e.g. TEB), whereas others which are relatively new (e.g. ACASA) have not been.

ACASA

ACASA (Pyles et al. 2003), developed by University of California – Davis, is a multilayer model that extends to 100 m above the canopy elements to ensure applicability of the turbulence assumptions. The canopy height considered is assumed to have the maximum eddy covariance (EC) tower height. In its current configuration, buildings and vegetation are assumed to be the same height. The model uses mass conservation with the absorbed available energy partitioned into sensible and latent heat flux densities. Energy balance closure is not forced, and the available energy partitioning is calculated using the Bowen ratio to ensure conservation of energy.

ACASA uses higher-order closure equations to estimate turbulent fluxes and profiles (Meyers and Paw U 1986, 1987) and to predict effects that higher-order turbulent kinetic and thermodynamic processes have on the surface microenvironment and associated fluxes of heat, moisture and momentum. These processes include turbulent production and dissipation of turbulence kinetic energy, turbulent vertical transport of heat, mass and momentum fluxes.

Multiple surface element angle classes (nine sunlit and one shaded) and direct as well as diffuse radiation absorption, reflection, transmission and emission are considered to estimate energy fluxes for each layer. For each angle class, the flux sources and sinks are estimated for each canopy element type (leaf, stem, building wall) and surface state (dry, wet, ice/snow-covered). Built surfaces and stems are assumed always to be dry.

Three types of surfaces are included in the model (building materials, vegetation and soil); their characteristics have to be specified to run the model. Building properties such as the values of thermal conductivity and emissivity, heat capacity of building material and the internal building temperature are adjustable, to allow different scenarios to be considered (i.e. changing roof albedo, introducing vegetated facades, etc.). Urban fluxes are calculated proportional to population density: the more people there are in an area, the more built surfaces are assumed to be influencing the flux sources and sinks.

As input the model requires meteorological information (air temperature, wind speed, specific or relative humidity, precipitation and air pressure), down-welling short and long wave radiation and carbon dioxide (CO_2) concentration. Inputs can be derived from surface measurements, or by meso-scale meteorological models, as well as morphological parameters to describe the surface.

ACASA has been coupled to the meso-scale Weather Research and Forecasting (WRF) Model to be the surface-layer scheme beneath the lowest sigma-layer. When used with WRF, the Mellor-Yamada-Janjic (Janjic 1994) Planetary Boundary Layer (PBL) scheme was used with the ACASA determined Turbulence Kinetic Energy (TKE) to calculate the PBL development. In the BRIDGE project, ACASA was run for the Firenze and Helsinki case studies (Marras et al. 2012).

SLUCM

One of the urban land surface schemes in the WRF meso-scale model couples the Noah land surface model (Chen & Dudhia 2001) and the SLUCM (Kusaka et al. 2001; Kusaka & Kimura 2004) through a tile approach. No interaction occurs between the two components, and the resulting outputs/fluxes are areally weighted as a function of the land cover fractions within each grid area. The model has been tested for numerous cities (e.g. Loridan et al. 2010; Loridan & Grimmond 2012), including London (Loridan et al. 2013).

The original model is described in detail by Kusaka et al. (2001) and Kusaka & Kimura (2004). For the vegetated part of the urban surface, the Chen & Dudhia (2001) Noah scheme is used. Loridan et al. (2010) modified SLUCM to improve its performance; see the discussion by Chen et al. (2011) which puts these changes in the context of other urban components. The model was evaluated in the urban land surface model comparison (Grimmond et al. 2010, 2011) by more than one group.

The model generally does not simulate sensible heat flux well. This is thought to be because of the lack of interaction between the surface tiles, which means that the excess turbulent sensible heat flux cannot be dissipated via evaporation (Loridan et al. 2010; Loridan and Grimmond 2012). Also, the model appears to have been originally optimized for radiation (Loridan & Grimmond 2012). Therefore changing coefficient values for the different parameters needs to be undertaken with caution when using the model for different scenarios, as in some cases it may result in the incorrect behaviour of the model.

SUEWS and LUMPS

SUEWS allows both the energy and water balances at the neighbourhood scale to be calculated. The model combines the urban evaporation–interception scheme of Grimmond & Oke (1991) with the urban water balance model of Grimmond et al. (1986). Recent developments (Järvi et al. 2011) explicitly aim to reduce the number of required input variables and to include more fully the energy and water exchange processes. In SUEWS particular attention is given to the surface conductances (or the inverse – resistances) with a Jarvis (1976) approach used in the Penman–Monteith equation (Penman 1948; Monteith 1965). Järvi et al. (2011) provide general coefficients for urban environments. The model has surface and subsurface soil moisture stores. The above ground surfaces include paved areas, roofs, evergreen trees and shrubs, deciduous trees and shrubs, grass and water. The vegetation can be irrigated and/or unirrigated. When the soil is saturated, excess water becomes surface runoff and/or leaves the bottom of the modelled soil layer, as deep soil runoff.

The model is forced with commonly measured meteorological variables (wind speed, relative humidity, air temperature, pressure, precipitation, shortwave irradiance). Surface characteristics are required for each model grid area, including the plan area fraction of each surface type, number of inhabitants, fraction of irrigated area using automatic sprinklers and internal hydrological connectivity (for example, based on elevation differences or pervious/impervious linkages or by piped network connectivity). The model undertakes calculations with a five minute to hourly time step. The results are then aggregated to daily, monthly and annual time periods. Thus, SUEWS can simulate or be applied to periods of (less than) a day to multiple years. For each time period, surface characteristics and meteorological forcing can change, or not, as appropriate for the simulation. Other recent developments include the addition of snow to the model so that it can more correctly simulate cold climate characteristics (Järvi et al. 2014). For a full description of the model see Järvi et al. (2011).

Within SUEWS, the simpler model LUMPS is included (Grimmond & Oke 2002). Both SUEWS and LUMPS treat the radiation (Offerle et al. 2003; Loridan et al. 2011) and storage heat flux (Grimmond et al. 1991; Grimmond & Oke 1999) in a common way. Evaluation of the full LUMPS model was conducted by Loridan et al. (2011) and SUEWS by Järvi et al. (2011). Both models underwent additional development for the BRIDGE project. LUMPS participated in the international urban land surface model comparison, whereas SUEWS did not; like ACASA it was not ready.

TEB

TEB (Masson 2000) simulates the energy and water exchanges between the city and the atmosphere. The most important processes that influence urban-atmosphere energy exchanges are taken into account in TEB, viz: radiative trapping and shadows resulting from the 3D geometry of a city; heat exchanges between the buildings and the environment; water interception and evaporation, and also changes in snow on roads and roofs (evaluated against Montreal data in Lemonsu et al. (2010)); and drag, heat and water turbulent exchanges between the urban canopy layer and the atmosphere.

The 3D shape of a city is parameterized using an idealized 2D canyon geometry while keeping the main features driving the radiative interactions and energy exchanges. Likewise, energy balance computations are carried out by azimuthal averaging solar and wind forcing in order to represent neighbourhoods with random-oriented urban canyons. For impact studies, a version of the model with specific canyon orientations is also available (Lemonsu et al. 2012). The air flow within urban canyons is solved by applying aerodynamic resistances and, in the latest version, by applying an original 1D vertical turbulence scheme that simulates the mean characteristics of the flow in the canyon, skipping unnecessary (and computationally expensive) details (Hamdi & Masson 2008). Generally the parameterization is designed to allow fast computations.

The Building Energy Model (BEM; Bueno et al. 2012) implemented in TEB considers a single thermal zone, a generic thermal mass to represent the thermal inertia of the indoor materials, the heat gains resulting from transmitted solar radiation and the internal sources of heat, infiltration and ventilation. The heat conduction through the envelope of the building is calculated using a finite difference method individually for each surface (roof, wall and floor). The morphological parameters of the TEB model are summarized in Table 9.1.

TABLE 9.1 TEB morphological parameters

Symbol	Description	Unit	Diagram and equations
Descriptive parameters of the city and the buildings (from urban data bases)			
A_{URB}	Urban area	m²	
A_{BLD}	Building plan area	m²	
A_{WALL}	Wall area	m²	
A_{WIN}	Window area	m²	
A_{MASS}	Indoor thermal mass area (walls, floors)	m²	
TEB input parameters			
λ_{BLD}	Building plan area density	–	$\lambda_{BLD} = \dfrac{A_{BLD}}{A_{URB}}$ $W\_o\_H = \dfrac{A_{WALL}}{A_{URB}}$
W_o_H	Wall-to-horizontal urban area ratio	–	
GR	Façade glazing ratio	–	$GR = \dfrac{A_{WIN}}{A_{WALL}}$
h_{BLD}	Building height	m	
h_{FLOOR}	Floor height	m	

LUCY

LUCY allows the anthropogenic heat flux (Q_F), and its spatial and temporal variability to be modelled (full details in Allen et al. 2011; Lindberg et al. 2013). The model includes the sensible heat released from transport or mobile sources, fixed or building sources and from people. LUCY calculates hourly fluxes and incorporates effects of the density of people across a city, monthly mean temperature, typical diurnal and day of week behaviour, details of the vehicle fleet and energy consumption for the country.

LUCY was run for all BRIDGE cities at the same spatial resolution using a consistent methodology. Here the model results are presented for one winter weekday (5 February 2008) and one summer weekday (16 July 2008) at 30 arc-second resolution (equivalent to 928 m x 465 m in London). Only urban grid cells as classified by CIESIN et al. (2004) are included in the analysis, and monthly mean temperatures from the Willmott et al. (2009) dataset are used. Two areas of London are modelled: the greater London area (GLA) and the central activity zone (CAZ). The extents of each model domain and the population within them are shown in Table 9.2.

Application and results

The performances of the models are evaluated here based on observations collected in Helsinki, Firenze and London (see details of the observations in Chapter 5). The widest range of models was run in Helsinki. Direct comparison is possible for some of the models but not all. The anthropogenic heat flux model, LUCY, is the only model that is run for all five cities (Athens, Gliwice, Helsinki, Firenze and London).

Helsinki

In Helsinki four surface energy balance models (ACASA, LUMPS, SUEWS and SLUCM) were compared with the EC observations undertaken at the Station for Measuring Ecosystems – Atmosphere Relations (SMEAR) III (see Chapter 5) Kumpula site (60°12′ N, 24°57′ E). The modelling groups ran their models for different periods. Here we consider the performance of four of the models for the period January to August 2008. The ACASA model was run with parameters indicated in Table 9.3 both offline (at the local-scale) and with WRF for a larger regional area. The SLUCM parameters and setup (see all details in Loridan and Grimmond 2012 used here are for Stage 4); all input parameters are optimized using the Multi Objective Shuffled Complex Evolution Metropolis (MOSCEM; Vrugt et al. 2003) algorithm. Thus this should be the best performance for this model.

Figure 9.1 shows the mean monthly diurnal pattern of the observed and modelled net all-wave radiation, sensible and latent heat fluxes for ACASA, LUMPS, SLUCM and SUEWS. All four models are able to reproduce the diurnal behaviour of the fluxes well (Figure 9.1). The mean statistics for the

TABLE 9.2 Modelled extents (lat, lon) used in LUCY and population density for each area

	London/GLA	Athens	Helsinki	Gliwice	Firenze	London/CAZ
lat_{min}	51.29	37.83	60.14	50.26	43.74	51.48
lat_{max}	51.70	38.13	60.30	50.33	43.81	51.54
lon_{min}	−0.50	23.60	24.80	18.60	11.19	−0.20
lon_{max}	0.30	23.80	25.10	18.74	11.31	−0.07
pop/km²	2922	4464	1700	1354	2770	8711
area (km²)	2865	559	310	110	93	107

TABLE 9.3 Input parameters used in ACASA runs in Firenze and Helsinki

Parameter	Firenze	Helsinki	References
Human population density (people m^{-2})	0.0036	0.0027	ISTAT 2009; Population Register Center of Finland 2009
Leaf Area Index (m^2 m^{-2})	0.5	1.5	This study
Canopy height (m)	36	31	Matese et al. 2009; Järvi et al. 2009
Interior building temperature (°C)	19	19	American Society of Heating, Refrigeration, and Air Conditioning Engineers 1992
Reflectivity (Visible)	0.2	0.2	Parker et al. 2000
Reflectivity (Near Infrared)	0.3	0.3	Parker et al. 2000
Thermal emissivity	0.9	0.9	Parker et al. 2000

models for each flux are presented in Table 9.4. For January ACASA overestimated the net all-wave radiation with mean bias error (*MBE*) of 12.7 W m^{-2}, whereas for SUEWS *MBE* was 0.4 W m^{-2} and for LUMPS 1.0 W m^{-2}.

Slight overestimation was simulated with SLUCM (*MBE* = −4.1 W m^{-2}). Three models were able to capture the behaviour of the winter time turbulent sensible heat flux (Q_H) well (except SLUCM) and differences between the models were very small, both by day- and by night-time. Turbulent latent heat flux (Q_E) on the other hand was underestimated by SUEWS, LUMPS and SLUCM (*MBE* = −12, −19.8 and -13.9 W m^{-2}) and overestimated by ACASA (*MBE* = 30.5 W m^{-2}).

Larger differences between the models were observed in August than in January. ACASA tends to overestimate net all-wave radiation ($Q\star$) and Q_H (*MBE* = 13.3 and 25.9 W m^{-2}), but is able to reproduce Q_E very well (*RMSE* = 24.3 W m^{-2} and *MBE* = −2.7 W m^{-2}). In contrast, both SUEWS and LUMPS underestimated $Q\star$ in the daytime, but were able to simulate nocturnal radiation levels correctly. SUEWS underestimated the daytime Q_H, but was able to predict the diurnal pattern in Q_E. LUMPS, in contrast, underestimates Q_E but simulates Q_H correctly. SLUCM underestimates $Q\star$ (*MBE* = −22.5 W m^{-2}) particularly at night-time, overestimates Q_H (*MBE* = −15.8 W m^{-2}) and clearly underestimates Q_E, similar to LUMPS. The conclusions are similar to those obtained in the Urban Project for Intercomparison of Land-surface Parameterization Schemes (PILPs) model comparison study (Grimmond et al. 2011); individual models have their strengths and weaknesses in modelling the different energy balance components.

When WRF-ACASA was used to calculate the fluxes for a 0.6 km by 0.6 km grid size inner domain (Falk et al. 2013), the instantaneous (30-minute) fluxes derived were compared for the entire month of June 2008. The Q_H and Q_E fluxes have similar magnitude to the daily maxima, indicating the effects of significant vegetation in the flux footprint with a Bowen ratio (β = Q_H/Q_E) of roughly 1. Generally, the model and observations had the same daily pattern. Typically the model errors were within the error range of the EC observations. The average peaks for both were 250 W m^{-2} for Q_H and 200 W m^{-2} for Q_E. WRF-ACASA captured heterogeneous forcing related to land cover differences across the urban landscape (Figure 9.2). Differences between night-time and day-time fields also indicate robust simulation of the diurnal cycle for the different land use types.

Firenze

Energy and mass fluxes were simulated in the Firenze city centre with an uncoupled or *in situ* version of ACASA for two years (2008 and 2010) using a 30-minute time step. Meteorological data (air temperature, relative humidity, precipitation and wind speed) collected at the Ximeniano Observatory

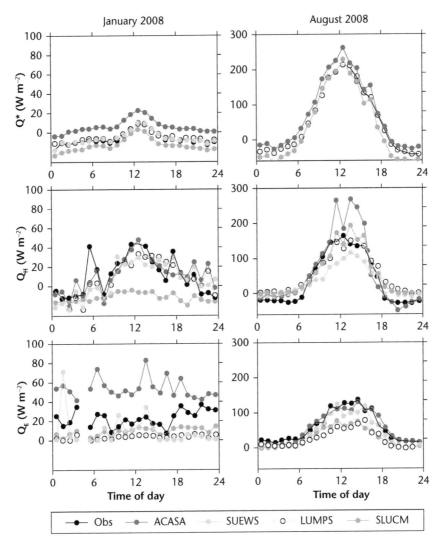

FIGURE 9.1 Median observed and simulated $Q^\star$, Q_H and Q_E in Helsinki using ACASA, SUEWS and LUMPS models for January and August 2008. Time is local time (UTC + 2 h)

(43°47′ N, 11°15′ E) and a constant CO_2 concentration value (388 parts per million (ppm), annual mean Mauna Loa Observatory, Hawaii, 2009–2010; NOAA 2012) were used to force the model. The morphological parameters related to vegetation (i.e. Leaf Area Index (LAI), maximum rate of Rubisco-mediated carboxylation V_{cmax}) are for broadleaf species that grow rapidly (large V_{cmax}, for instance) in urban areas. However, the city centre of Firenze has very little vegetation, so a small LAI value was set for this city (Table 9.3). Note the model does not use the percentage of vegetation surface cover, so to control the vegetation amount the LAI is changed. The soil type class corresponding to concrete surfaces was used for simulations.

Simulated energy fluxes were compared with EC flux measurements collected at the Ximeniano tower for the periods January–April 2008 and July–September 2010. ACASA reproduces diurnal variation in Q_H flux, but overestimates the midday fluxes by more than one standard deviation of the observations (e.g. April 2008 period; Figure 9.3). The scarcity of vegetation (< 5 per cent; Vaccari et al.

TABLE 9.4 Observed (mean) and model performance evaluation statistics (*RMSE*; W m⁻² and *MBE*, W m⁻²) for four urban land-surface models. The model runs and observations are for Helsinki in January and August 2008. The number of 30 minute data points used in the statistics are indicated by *N*

		Q^*		Q_H		Q_E	
		Mean	*N*	*Mean*	*N*	*Mean*	*N*
Observed	Jan	−20.9	720	17.6	187	23.2	58
	Aug	69.4	738	37.3	341	63.0	311
Modelled		*RMSE*	*MBE*	*RMSE*	*MBE*	*RMSE*	*MBE*
ACASA	Jan	7.4	12.7	40.4	−14.3	13.7	30.5
	Aug	39.6	13.3	53.1	25.9	24.3	−2.7
SUEWS	Jan	2.9	0.4	24.4	−12.5	16.5	−12.0
	Aug	5.1	−4.7	31.6	−4.6	29.8	−14.0
LUMPS	Jan	2.9	1.0	24.1	−11.8	1.9	−19.8
	Aug	5.1	−4.7	37.9	15.8	19.7	−33.7
SLUCM	Jan	7.0	−4.5	10.4	−30.1	4.2	−13.9
	Aug	41.9	−22.5	29.1	15.8	17.1	−32.9

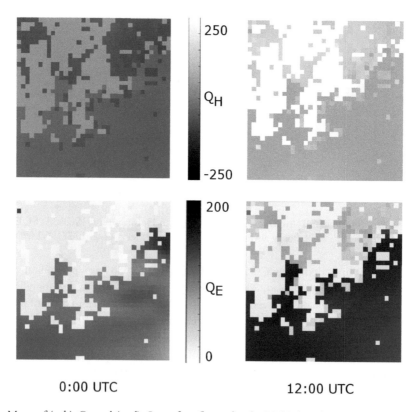

FIGURE 9.2 Maps of (a, b) Q_H and (c, d) Q_E surface fluxes for the Helsinki urban area modelled with WRF-ACASA (domain 5, Δx = 0.6 km, 18.6 × 18.6 km) for (a, c) 00:00 and (b, d) 12:00 UTC on 15 July 2008 (modified from Falk et al. 2013)

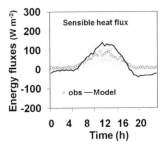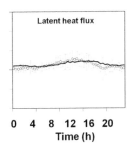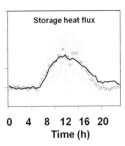

FIGURE 9.3 Mean diurnal sensible, latent and storage fluxes in April 2008 observed and ACASA simulated for Firenze (modified from Marras et al. 2012). Observations are 33 m above ground level and mean building height of the surrounding area is 25 m. The model is forced with data at the same height as the observations. Time is local time.

2013) in the area results in small latent heat fluxes (Q_E), which ACASA is able to capture. The model evaluation statistics for April 2008 (N = 1295, 30-minute data periods) have a RMSE of 32 W m^{-2} and a *MBE* of 6 W m^{-2}. The observed net storage heat flux (ΔQ_S) was estimated as residual from the energy balance equation assuming the anthropogenic heat flux (Q_F) is included in the other terms or is negligible. There is good agreement between simulated and estimated ΔQ_S (Figure 9.3), with a *RMSE* of 60 W m^{-2} and a *MBE* of 3 W m^{-2}.

Fluxes at the regional scale were also calculated by running the coupled model WRF-ACASA at 200-m horizontal grid for a 5 km x 5 km domain (Figure 9.4). The model setup involved six nested grids from 48 km down to 200 m grid spacing, with a nesting ratio of 3. National Centers for Environmental Prediction/National Center for Atmospheric Research (NCEP/NCAR) Reanalysis (NNRP) data were used for WRF initial and boundary conditions for the coupled WRF-ACASA simulation. Modelled surface fluxes (Q_H, Q_E and ΔQ_S) have the expected patterns relative to the land cover (Figure 9.4) and surface types within the domain. Variations between diurnal and nocturnal flux values are shown in Figure 9.4, due to land use and local meteorological effects. Simulated values were within the range of observations.

London

Performance of SLUCM/Noah was evaluated both offline and within WRF (WRF-SLUCM/Noah for 3 June 2010; Loridan et al. 2013), for a period when a high pressure system was over the UK giving clear-sky conditions. Two sets of EC tower observations located at King's College London (KCL, 51.511° N, 0.116° W) were used to evaluate the model performance (see Kotthaus and Grimmond 2012, 2013a,b and Loridan et al. 2013 for measurement details; also described in Chapter 5).

The SLUCM/Noah offline application uses local scale observations (incoming shortwave and long-wave radiation, air temperature, relative humidity, wind components and pressure) to force the model. When the scheme is linked to WRF-SLUCM/Noah (i.e. online) the forcing fields are simulated by WRF (see Loridan et al. 2013 for details of the modelling setup). For both offline and online runs the modelled fluxes are evaluated and the impact of the land-surface characterization assessed. For the online simulations the modelled meteorological variables can also be assessed. The spatial extent of the observed EC and radiation source area are smaller than the 1 km^2 online grid cell used in WRF.

The diurnal evolution of the observed surface energy balance, the offline and online model fluxes are shown in Loridan et al.'s (2013) Figures 4 and 5. The model was run in Helsinki with a wide variety

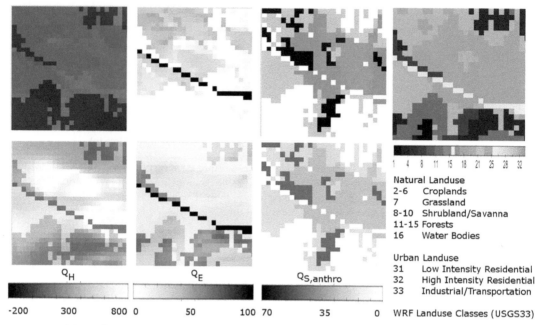

FIGURE 9.4 Maps of Q_H, Q_E and storage heat ($Q_{s\,anthro}$) surface fluxes at (a, b, c) 5 am and (d, e, f) 12 noon local time (a, d) turbulent sensible heat flux, (c, f) storage heat flux in impervious surface elements (buildings, roads, etc.), (b, e) latent heat flux for Firenze on 15 April 2008 (modified from Marras et al. 2012). These are modelled with WRF-ACASA at 200 m × 200 m resolution at the lowest WRF-sigma layer (approximately 50 m above the surface). (g) Land use map (WRF-ACASA) for Firenze, Italy. Dark grey colours indicate natural land cover, light grey colours urban land cover classes, white indicates water bodies (WRF 33 Land-use Classes, USGS33)

of land surface parameter values. These include the default parameter values for WRF V3.2.1 (termed "stage 1"), with high values for the albedo (0.2) for the roof, road and wall. This results in a clear overestimation of reflected shortwave K↑ (see Loridan et al.'s 2013, Figure 4a; RMSE =25 W m^{-2} and MBE of 12 W m^{-2} see their Table 4). At stage 5, when the parameter values are changed to the urban zone for energy partitioning (UZE) values recommended by Loridan & Grimmond (2012), appropriate for the measurement site (in this case high density, HD), the model performance statistics are improved (Table 9.5). Similarly, the modelled daytime L↑ is overestimated at stage 1, while stage 5 (UZE) yields results in better agreement with observations (*RMSE* reduced from 37 W m^{-2} to 11 W m^{-2}, Table 9.5). The daytime Q⋆ is underestimated at stage 1 (*RMSE* ≈ 59 W m^{-2}, *MBE* ≈ −43 W m^{-2}), but this bias is reduced by using UZE parameters (*RMSE* ≈14 W m^{-2} *MBE*=−6 W m^{-2}). However, for both (stages 1 and 5) peak daytime Q_H is overestimated, whereas evening/early morning values are underestimated. With less Q⋆ at the surface, the stage 1 parameters cause less daytime Q_H release (*RMSE* = 62 W m^{-2}). Although the underestimated Q_E (*MBE* ≈ −19 W m^{-2}) is improved with UZE parameters (Table 9.5), the variability in the Q_E observations means the statistics are not very robust.

Anthropogenic heat flux in all five cities – LUCY

LUCY was used to model diurnal variations of Q_F for a winter and a summer weekday for the five cities (Figure 9.5). Results show that the central parts of London (CAZ) have the highest values

TABLE 9.5 Performance of Noah/SLUCM in London on 3 June 2010 (data from Loridan et al. 2013, Stage 5b) compared to EC flux observations. Land surface parameters for model runs are UZE values

$W\ m^{-2}$	Offline		Online	
	RMSE	MBE	RMSE	MBE
$K\uparrow$	14	−8	14	−2
$L\uparrow$	11	8	15	0
$Q\star$	14	6	29	−8
Q_H	83	27	83	5
Q_E	14	−6	16	−5

of Q_F, during both winter and summer. The main reason for this is a high population density in combination with relatively high energy consumption which affects the heat released from buildings in the model. Examining the fluxes across the full extent of the Greater London Authority, however, yields a mean Q_F which is low because of lower population densities in the suburbs compared to the other cities examined. In winter, Helsinki has a relatively large release of Q_F even though the population density is low (Table 9.2). This is because of the low winter temperatures which result in a high proportion of the yearly energy consumption used for heating during the cold winter months. In contrast, during summer Q_F for Helsinki is the second lowest because little energy is needed to heat or cool buildings. Athens has Q_F fluxes very close to those of the dense London CAZ in summer given the high proportion of the energy used to cool buildings during the hot summer. In contrast, cities in UK use very little energy to cool buildings during summer (see Figure 4 in Lindberg et al. 2013), hence energy released by buildings as well as the total Q_F is much lower in summer compared to winter. Q_F in Firenze is relatively low because the relatively low population density. The difference in Q_F between summer and winter is very small as similar amounts of energy are used to heat buildings in winter as well as cool buildings during summer. Gliwice has the lowest release of Q_F, mainly because the population density is so low.

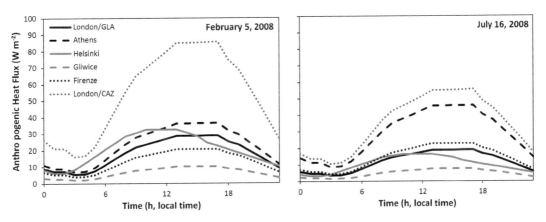

FIGURE 9.5 Diurnal variations of average anthropogenic heat flux for a winter and summer weekday for the five BRIDGE cities in Europe, 2008

Conclusions

Using the EC observations (see Chapter 5 for details) collected as part of BRIDGE, it has been possible to test the performance of several urban land surface models to simulate the urban energy balance exchanges. The models used here include models included in the international urban land surface model comparison (TEB, Noah/SLUCM, LUMPS) and two models that have been developed further within this project (ACASA, SUEWS).

The international urban land surface model comparison concluded that no model had the best overall capability for modelling the surface energy balance fluxes within an urban area (Grimmond et al. 2010, 2011). The results here lead us to similar conclusions. No individual model is best or worst across the cities in which they were tested for all of the fluxes considered. All of the models have benefited from improvements and additional capabilities added through the BRIDGE project (e.g. Järvi et al. 2011; Loridan et al. 2011; Falk et al. 2013; Järvi et al. 2014). This is an important contribution to the evolution of urban land surface modelling.

The model runs demonstrate the importance not only of model physics but also the surface parameters used for the model runs. The need to ensure that these are considered as a whole to understand their relative importance and their suitability for a given site is critical when land surface modifications are to be considered. If the models are not capable of modelling current conditions and/or responding appropriately to parameter changes, then simulations for future scenarios and different interventions could be very misleading.

Here the SLUCM model is documented to systematically have a problem with the turbulent sensible heat fluxes, notably the ability to transfer heat between sensible and latent heat exchanges. The performance was improved though when appropriate combined model parameters are used (i.e. UZE rather than default WRF values). This has implications that need to be taken into account when modelling surface changes with SLUCM, for example scenarios such as planting trees. Although the ACASA model results are promising, further refinements in the model parameterization are needed to improve the impact of urban surface characteristics on simulated energy fluxes.

References

Allen, L., F. Lindberg & C.S.B. Grimmond (2011). Global to city scale model for anthropogenic heat flux. *International Journal of Climatology*, 31, 1990–2005.

American Society of Heating, Refrigeration, and Air Conditioning Engineers (ASHRAE) (1992). Report Number 155–1192 *Thermal Environment Conditions for Human Occupancy*. Atlanta, GA USA.

Bueno, B., G. Pigeon, L. Norford, K. Zibouche & C. Marchadier (2012). Development and evaluation of a building energy model integrated in the TEB scheme. *Geoscientific Model Development*, 5, 433–448.

Chen, F. & J. Dudhia (2001). Coupling an advanced land surface–hydrology model with the Penn State–NCAR MM5 modelling system. Part I: Model implementation and sensitivity. *Monthly Weather Review,* 129, 569–585.

Chen F., H. Kusaka, R. Bornstein, J. Ching, C.S.B. Grimmond, S. Grossman-Clarke, T. Loridan, K.W. Manning, A. Martilli, S. Miao, D. Sailor, F.P. Salamanca, H. Taha, M. Tewari, X. Wang, A.A. Wyszogrodzki & C. Zhang (2011). The integrated WRF/urban modeling system: development, evaluation, & applications to urban environmental problems, *International Journal of Climatology*, 31, 273–288.

Chrysoulakis, N., M. Lopes, R. San José, C.S.B. Grimmond, M.B. Jones, V. Magliulo, J.E.M. Klostermann, A. Synnefa, Z. Mitraka, E. Castro, A. González, R, Vogt, T. Vesala, D. Spano, G.Pigeon, P. Freer-Smith, T. Staszewski, N. Hodges, G. Mills & C. Cartalis (2013). Sustainable urban metabolism as a link between biophysical sciences and urban planning: the BRIDGE project. *Landscape and Urban Planning*, 112, 100–117.

CIESIN, Columbia University; International Food Policy Research Institute (IFPRI), the World Bank; and Centro Internacional de Agricultura Tropical (CIAT) (2004). Global Rural–Urban Mapping Project (GRUMP), Alpha Version: Urban Extents. Palisades, NY: CIESIN, Columbia University. <Online>. Available from: http:// sedac.ciesin.columbia.edu/gpw, Accessed 30 July 2009.

Falk, M., R.D. Pyles, S. Marras, D. Spano, K.T. Paw, U.R.L. & Snyder (2013). Regional simulation of urban metabolism in Helsinki, Finland in 2008. (under review).

Grimmond C.S.B., M. Best, J. Barlow, A.J. Arnfield J.-J. Baik, S. Belcher, M. Bruse, I. Calmet, F. Chen, P. Clark, A. Dandou, E. Erell, K. Fortuniak, R. Hamdi, M. Kanda, T. Kawai, H. Kondo, S. Krayenhoff, S.H. Lee, S.-B. Limor, A. Martilli, V. Masson, S. Miao, G. Mills, R. Moriwaki, K. Oleson, A. Porson, U. Sievers, M. Tombrou, J. Voogt, & T. Williamson (2009). Urban surface energy balance models: model characteristics & methodology for a comparison study. In *Urbanization of Meteorological & Air Quality Models*, (eds) Baklanov, A., C.S.B. Grimmond, A. Mahura & M. Athanassiadou Springer-Verlag, ISBN: 978–3–642–00297–7, 97–123.

Grimmond, C.S.B., M. Blackett, M. Best, J. Barlow, J.J. Baik, S. Belcher, S.I. Bohnenstengel, I. Calmet, F. Chen, A. Dandou, K. Fortuniak, M.L. Gouvea, R. Hamdi, M. Hendry, T. Kawai, Y. Kawamoto, H. Kondo, E.S. Krayenhoff, S.H. Lee, T. Loridan, A. Martilli, V. Masson, S. Miao, K. Oleson, G. Pigeon, A. Porson, Y.H. Ryu, F. Salamanca, G.J. Steeneveld, M. Tombrou, J. Voogt, D. Young & N. Zhang (2010). The International Urban Energy Balance Models Comparison Project: first results from Phase 1. *Journal of Applied Meteorology & Climatology*, 49, 1268–1292, doi: 10.1175/2010JAMC2354.1

Grimmond, C.S.B., M. Blackett, M.J. Best, J.-J. Baik, S.E. Belcher, J. Beringer, S.I. Bohnenstengel, I. Calmet, F. Chen, A. Coutts, A. Dandou, K. Fortuniak, M.L. Gouvea, R. Hamdi, M. Hendry, M. Kanda, T. Kawai, Y. Kawamoto, H. Kondo, E.S. Krayenhoff, S.-H. Lee, T. Loridan, A. Martilli, V. Masson S. Miao, K. Oleson, R. Ooka, G. Pigeon, A. Porson, Y.-H. Ryu, F. Salamanca, G.-J. Steeneveld, M. Tombrou, J.A. Voogt, D. Young & N. Zhang (2011). Initial Results from Phase 2 of the International Urban Energy Balance Comparison Project, *International Journal of Climatology*, 31, 244–272, doi: 10.1002/joc.2227

Grimmond, C.S.B., H.A. Cleugh & T.R. Oke (1991). An objective urban heat storage model & its comparison with other schemes. *Atmospheric Environment*, 25B, 31–326.

Grimmond, C.S.B. & T.R. Oke (1991). An evaporation-interception model for urban areas. *Water Resources Research*, 27, 1739–1755.

Grimmond, C.S.B. & T.R. Oke (1999). Heat storage in urban areas: observations & evaluation of a simple model. *Journal of Applied Meteorology*, 38, 922–940.

Grimmond, C.S.B. & T.R. Oke (2002). Turbulent heat fluxes in urban areas: Observations & local-scale urban meteorological parameterization scheme (LUMPS). *Journal of Applied Meteorology,* 41, 792–810.

Grimmond, C.S.B., T.R. Oke & D.G. Steyn (1986). Urban water balance I: a model for daily totals. *Water Resources Research*, 22, 1397–1403.

Hamdi, R. & V. Masson (2008). Inclusion of a drag approach in the town energy balance (TEB) scheme: offline 1-d validation in a street canyon. *Journal of Applied Meteorology & Climatology*, 47, 2627–2644.

ISTAT (2009) http://demo.istat.it/index_e.html. Retrieved 15 September 2013

Janjic, Z.I. (1994). The step–mountain eta coordinate model: further developments of the convection, viscous sublayer and turbulence closure schemes. *Monthly Weather Review,* 122, 927–945.

Järvi, L., C.S.B. Grimmond & A. Christen (2011). The surface urban energy and water balance scheme (SUEWS): evaluation in Los Angeles and Vancouver, *Journal of Hydrology*, 411, 219–237.

Järvi, L., C.S.B. Grimmond, M. Taka, A. Nordbo, H. Setälä & I. Strachan (2014). Modelling of the energy and water balance in cold climate cities using the Surface Urban Energy and Water balance Scheme (SUEWS), *Geoscientific Model Development Discussion*, 7, 1063–1114.

Järvi, L., C.S.B. Grimmond, M. Taka, H. Setälä, A. Nordbo & I.B. Strachan (2013). Development of the Surface Urban Energy and Water balance Scheme (SUEWS) for cold climate cities. Geoscientific Model Development (GMD) MS No.: gmd-2013–163. http://www.geosci-model-dev-discuss.net/7/1063/2014/gmdd-7–1063–2014.pdf

Järvi, L., H. Hannuniemi, T. Hussein, H. Junninen, P.P. Aalto, R. Hillamo, T. Mäkelä, P. Keronen, E. Siivola, T. Vesala & Kulmala, M. (2009). The urban measurement station SMEAR III: continuous monitoring of air pollution and surface-atmosphere interactions in Helsinki, Finland. *Boreal Environment Research*, 14 (Suppl. A), 86–109.

Jarvis, P.G. (1976). The interpretation of the variations in leaf water potential and stomatal conductance found in canopies in the field. *Philosophical Transactions of the Royal Society London*, Ser. B. 273, 593–610.

Kotthaus, S. & C.S.B. Grimmond (2012). Identification of micro-scale anthropogenic CO_2, heat and moisture sources - processing eddy covariance fluxes for a dense urban environment. *Atmospheric Environment*, 57, 301–316.

Kotthaus, S. & C.S.B. Grimmond (2013a). Energy exchange in a dense urban environment – Part I: temporal variability of long-term observations in central London. *Urban Climate*, doi: 10.1016/j.uclim.2013.10.002

Kotthaus, S. & C.S.B. Grimmond (2013b). Energy exchange in a dense urban environment – Part II: impact of spatial heterogeneity of the surface. *Urban Climate*, 10.1016/j.uclim.2013.10.001

Kusaka, H. & F. Kimura (2004). Thermal effects of urban canyon structure on the nocturnal heat island: numerical experiment using a mesoscale model coupled with an urban canopy model. *Journal of Applied Meteorology and Climatolgoy*, 43, 1899–1910.

Kusaka, H., H. Kondo, Y. Kikegawa & F. Kimura (2001). A simple single layer urban canopy model for atmospheric models: comparison with multi-layer and slab models. *Boundary-Layer Meteorology*, 101, 329–358.

Lemonsu, A., S. Bélair, J. Mailhot & S. Leroyer (2010). Evaluation of the Town Energy Balance model in cold and snowy conditions during the Montreal Urban Snow Experiment 2005. *Journal of Applied Meteorology & Climatology*, 49, 346–362.

Lemonsu, A., V. Masson, L. Shashua-Bar, E. Erell & D. Pearlmutter (2012). Inclusion of vegetation in the Town Energy Balance model for modelling urban green areas. *Geoscientific Model Development*, 5, 1377–1393.

Lindberg F., C.S.B. Grimmond, N. Yogeswaran, S. Kotthaus & L. Allen (2013). Impact of city changes and weather on anthropogenic heat flux in Europe 1995–2015. *Urban Climate*, http://dx.doi.org/10.1016/j.uclim.2013.03.002

Loridan, T. & C.S.B. Grimmond (2012). Multi-site evaluation of an urban land-surface model: intra-urban heterogeneity, seasonality and parameter complexity requirements. *Quarterly Journal of the Royal Meteorological Society*, 138, 1094–1113, doi: 10.1002/qj.963

Loridan, T., C.S.B. Grimmond, S. Grossman-Clarke, F. Chen, M. Tewari, K. Manning, A. Martilli, H. Kusaka & M. Best (2010). Trade-offs & responsiveness of the single-layer urban canopy parameterization in WRF: an offline evaluation using the MOSCEM optimization algorithm & field observations. *Quarterly Journal of the Royal Meteorological Society*, 136: 997–1019. doi: 10.1002/qj.614

Loridan T., C.S.B. Grimmond, B.D. Offerle, D.T. Young, T. Smith, L. Järvi & Lindberg, F. (2011). Local-Scale Urban Meteorological Parameterization Scheme (LUMPS): longwave radiation parameterization & seasonality related developments. *Journal of Applied Meteorology & Climatology*, 50, 185–202.

Loridan T., F. Lindberg, O. Jorba, S. Kotthaus, S. Grossman-Clarke & C.S.B. Grimmond (2013). High resolution simulation of surface heat flux variability across central London with Urban Zones for Energy partitioning. *Boundary Layer Meteorology*, 147, 493–523.

Marras, S., R.D. Pyles, M. Falk, M. Casula, R.L. Snyder, K.T. Paw U, & D. Spano (2012). Urban fluxes estimation at local and regional scale by ACASA model. *ICUC8 – 8th International Conference on Urban Climates*, 6–10 August, 2012, UCD, Dublin, Ireland.

Masson, V. (2000). A physically-based scheme for the urban energy budget in atmospheric models. *Boundary-Layer Meteorology*, 94, 357–397.

Matese, A., B. Gioli, F.P. Vaccari, A. Zaldei & F. Miglietta (2009). CO_2 emissions of the city center of Firenze, Italy: measurement, evaluation and source partitioning. *Journal of Applied Meteorology & Climatology*, 48, 1940–1947.

Meyers T.P. & K.T. Paw U (1986). Testing of a higher-order closure model for modelling airflow within and above plant canopies. *Boundary-Layer Meteorology*, 31, 297–311.

Meyers T.P. & K.T. Paw U (1987). Modelling the plant canopy micrometeorology with higher-order closure principles. *Agricultural and Forest Meteorology*, 41, 143–163.

Monteith, J.L. (1965). Evaporation and environment. *Symposia of the Society for Experimental Biology*, 19, 205–224.

NOAA (2012). ftp://ftp.cmdl.noaa.gov/ccg/co2/trends/co2_annmean_mlo.txt (Accessed 29 October 2012).

Offerle, B.D., C.S.B. Grimmond & T.R. Oke (2003). Parameterization of net all-wave radiation for urban areas. *Journal of Applied Meteorology*, 42, 1157–1173.

Parker, D.S., J.E.R. McIlvaine, S.F. Barkaszi, D.J. Beal & M.T. Anello (2000). *Laboratory Testing of the Reflectance Properties of Roofing Material*. FSEC-CR-670-00. Florida Solar Energy Center, Cocoa, FL.

Penman, H.L. (1948). Natural evaporation from open water, bare soil and grass. *Proceedings of the Royal Society*, Ser. A. 193, 120–145.

Population Register Center of Finland (2009) Väestötietojärjestelmä Rekisteritilanne 31.8.2013 (in Finnish and Swedish). http://vrk.fi/default.aspx?docid=7675&site=3&id=0 (retrieved 15 September 2013).

Pyles, R.D., B.C. Weare, K.T. Paw U & W. Gustafson (2003). Coupling between the University of California, Davis, Advanced Canopy-Atmosphere-Soil Algorithm (ACASA) and MM5: preliminary results for July 1998 for Western –North America. *Journal of Applied Meteorology*, 42, 557–569.

Vaccari, F.P., B. Gioli, P. Toscano & C. Perrone (2013). Carbon dioxide balance assessment of the city of Florence (Italy), and implications for urban planning. *Landscape and Urban Planning*, 120, 138–146.

Vrugt, J.A., H.V. Gupta, L.A. Bastidas, W. Bouten & S. Sorooshian (2003). Effective and efficient algorithm for multiobjective optimization of hydrological models. *Water Resources Research*, 39 (12), 14, doi: 10.1029/2002WR001746

Willmott, C.J., K. Matsuura & D.R. Legates (2009). Terrestrial air temperature: 1900–2008 gridded monthly time series (1900–2008) (V. 2.01), <Online>. Available from: http://climate.geog.udel.edu/~climate/ html_pages/ download.html#lw_temp> (Accessed 15 January 2011).

10

URBAN WATER BALANCE AND HYDROLOGY MODELS TO SUPPORT SUSTAINABLE URBAN PLANNING

Eddy J. Moors[1], C.S.B. Grimmond[2], Ab Veldhuizen[1], Leena Järvi[3] and Frank van der Bolt[1]

[1]ALTERRA - WAGENINGEN UR, [2]UNIVERSITY OF READING AND [3]UNIVERSITY OF HELSINKI

Introduction

The urban water balance ensures conservation of mass of water in the same way the energy balance requires conservation of energy (Chapter 4 for details). By considering all the exchanges in the urban water balance, insight can be gained into the dynamic processes and feedback mechanisms in the urban environment, providing important insights into urban sustainability.

To create a sustainable urban area requires a coherent strategy that applies planning/design tools at the appropriate scale and ensures that actions at one scale are not counteracted at another. These strategies need to apply not only to the built area, but also to the surrounding area that is intertwined. To integrate these sustainability principles into urban planning strategies at the level of land use (1:5000–10,000; neighbourhood to settlement) and master plans (1:500–1000; building block), knowledge of the urban water balance is needed as guidance for innovative planning and design on a more detailed level. In addition, an accurate representation of the urban water balance through modelling is imperative for the assessment of future sustainable urban water management practices, realistic simulation of urban surface processes and for predicting the effects of climate change.

In urban areas the water system often consists of two separate parts that strongly interact: the semi-natural surface and groundwater system and the man-made sewer and supply systems. Knowledge of the semi-natural surface and groundwater system is especially important at the level of neighbourhood and master plans, while the man-made sewer and supply systems become essential at the more detailed level.

As this book aims to support planning processes at the scale of city to neighbourhoods and master plans, the emphasis in this chapter is on hydrological models rather than the more detailed hydraulic models needed to design an urban sewer system. After a short review of the available types of urban water balance models, two urban water balance models used in the BRIDGE project (Chrysoulakis et al. 2013) SIMGRO and Surface Urban Energy and Water Balance Scheme (SUEWS) are presented. These physically based hydrological models for urban spatial planning are applied in two BRIDGE case study cities, London and Helsinki. The energy and water balances are linked by the evaporative or latent heat flux terms (see Chapter 4 for details). The urban land surface models that determine the surface energy balance fluxes were tested independently, as presented in Chapter 9.

Overview of urban water balance models

In the literature three general types of urban water balance models exist with varying degrees of complexity and spatial extent. Each is described below.

Mass balance models

Models based on the mass balance are largely used to determine the urban water balance for urban hydrology and water management applications. They consider both natural and anthropogenic hydrological systems through the use of a number of empirical relations to determine the fluxes and storage of the urban system for the desired spatial and temporal scale. The latter depends on data availability and resolution (typically daily data are used). These models consider the inputs, outputs, flows and stores of the water balance.

Grimmond & Oke (1986) describe such a model and evaluate it using observations from Vancouver, Canada. That model has been used to study urban irrigation and the urban water link to the energy budget via evapotranspiration. Two urban water balance models developed in Australia for the assessment of water management techniques are Aquacycle (Mitchell et al. 2001) and the Urban Volume and Quality (UVQ) model (Mitchell & Diaper 2005). Aquacycle contains options to apply water management techniques to the urban water balance and was evaluated using data from Woden Valley, Canberra, Australia (Mitchell et al. 2003). UVQ is essentially an expanded version of Aquacycle with the added ability to model contaminant fluxes (Diaper & Mitchell 2007) for a number of cities around the world (Wolf et al. 2007). Site specific input values are required to calibrate and run the models with three nested spatial scales in each (unit block property), cluster (neighbourhood) and the study area as a whole (Wolf et al. 2007). Unlike the Urban Water Balance model of Grimmond et al. (1986), there is less focus on required meteorological data, with only daily precipitation and potential evapotranspiration values needed.

Surface atmosphere transfer schemes

Secondly, urban parameterization schemes used in global and meso-scale numerical weather models are available; examples include the Urban Hydrological Element (UHE) model (Berthier et al. 2004, 2006); the urbanized Submesoscale Soil Model (SM2-U) (Dupont et al. 2006); the combined Town Energy Balance and Interaction Soil-Biosphere-Atmosphere scheme (TEB-806 ISBA) (Lemonsu et al. 2007); and the SUEWS (Järvi et al. 2011). Each scheme differs in complexity and focus, but in essence all are formed of a number of surface and subsurface layers and model the surface water balance using inputs from a numerical atmospheric model (typically net radiation and precipitation) and generate outputs for use in the next model time step (evapotranspiration). All the schemes presented use no, or very simple, drainage networks and have mixed land uses (urban and natural surface types), each of which have individual surface and hydrologic properties weighted by their relative areal coverage of a particular grid box. Unlike the dedicated urban water balance models, these parameterizations focus only on the external water system.

The UHE model (Berthier et al. 2004, 2006) focuses on simulating storm water runoff and soil infiltration within the urban water balance. Two main layers are used: a surface layer with three possible land cover types (natural, paved and building roof) and a soil layer (subdivided into an upper and lower sub-layer for infiltration purposes), which takes the form of a fine mesh grid. In addition to these layers, there is a storm water drainage system which is represented as a trench collecting all available runoff as well as seepage from soil water (this acts in both directions depending on soil moisture conditions). The original version (Berthier et al. 2004) considered evapotranspiration (and infiltration) by applying a 'mixed' boundary condition on modelled soil moisture, observed rainfall and potential evapotranspiration. Further developments included a dual evapotranspiration scheme based on the Penman-Monteith-Rutter-Shuttleworth equation (Grimmond & Oke 1991) for paved and roof surfaces and a scheme based on Feddes et al. (1988) to calculate the potential evaporation and transpiration for the natural surface types (Berthier et al. 2006). Model evaluation was done using site specific parameters and observed meteorological and hydrological variables from the Reze field site, Nantes.

SM2-U is a mesoscale model surface parameterization scheme (Dupont et al. 2006) formed of four levels (lower atmosphere, surface layer, root zone and deep soil) and a rudimentary drainage network as used in Berthier et al. (2004). It is an extension of Noilhan & Planton's (1989) ISBA scheme with the addition of four artificial urban surface types (building roofs, paved surfaces, vegetation elements over paved surfaces and paved surfaces below vegetation), each with their own surface properties (Dupont et al. 2006). The scheme was evaluated with data from three measurement sites, two rural and the suburban site at Reze, which allowed comparison of the runoff running through the drainage network with the original version of UHE (Berthier et al. 2004). It was concluded that the SM2-U scheme performed well annually and in summer storm events in comparison to UHE but was poor at simulating winter storms as infiltration to and from the drainage network was not modelled.

Lemonsu et al. (2007) also used ISBA but with Masson's (2000) Town Energy Balance (TEB) urban surface parameterization scheme. The scheme has three layers (the surface and two soil layers) and four surface types, the three ISBA vegetation surfaces (bare soil, soil between vegetation and vegetation) and an urban surface which in TEB is modelled as a three dimensional urban canyon. An off-line simulation of the TEB-ISBA water balance was undertaken using meteorological data to force the model and input parameters from the literature relevant to a suburban area. The results were compared with data from the Reze study area. Discrepancies between modelled and observed runoff suggested further work on the parameterization of surface infiltration through roads was needed (Lemonsu et al. 2007).

The fourth surface parameterization scheme, SUEWS, simulates the urban water balance components including irrigation, surface runoff, infiltration, evaporation and storage at a local or neighbourhood scale (Järvi et al. 2011). It consists of three layers (the surface and two soil layers) and seven surface types (paved, buildings, evergreen and deciduous trees/shrubs, irrigated and non-irrigated grass and water). SUEWS does not have separate tiles for urban and vegetative surface fractions, but rather dynamic interaction between the different surface types is allowed. The model has been successfully run for Los Angeles, USA, and Vancouver, Canada. As its performance in cold climate cities in wintertime was poor due to the inadequate description of snow, further development has been undertaken (Järvi et al. 2014).

Hydrological models

Hydrological models are typically composed of two parts: a surface scheme and a hydrological flow model (described as a natural, anthropogenic or a combination drainage system often represented by a combination of a surface and groundwater module) that moves water through the study catchment. In Figure 10.1 an example of such a model configuration and the processes involved is depicted. Examples of models of this type include: Semi-Urbanised Runoff Flow (SURF) (Rodriguez et al. 2000); the Water and Energy transfer Processes (WEP) model (Jia et al. 2001); the Urban-Runoff Branching Structure Model (URBS-MO) (Rodriguez et al. 2008); the SIMGRO model (van Walsum & Veldhuizen 2011); and the Urban FORest Effects-Hydrology (UFORE-Hydro) model (Wang et al. 2008).

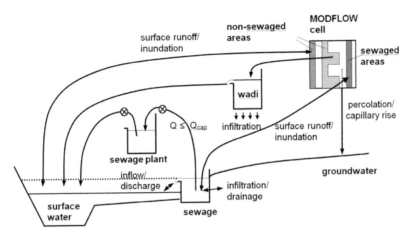

FIGURE 10.1 The urban water cycle as represented in the SIMGRO model (van Walsum & Veldhuizen 2011)

These models are typically used to study sewer and drain performance during rainfall events (Rodriguez et al. 2008), assessing impacts of runoff pollution (Rodriguez et al. 2000), simulation of the effects of urban areas on natural catchment flows and flooding (Jia et al. 2001) and the impact of urban trees and canopy interception on runoff and evapotranspiration (Wang et al. 2008; van Walsum & Veldhuizen 2011). The data inputs and parameters required for these models include physical land cover properties (e.g. fraction of impervious land cover and aerodynamic roughness length), surface and subsurface hydrologic properties (e.g. soil storage capacity), water use information (e.g. mean water use), initial conditions in terms of storage in the study area (e.g. soil moisture) and a range of averaged (period of choice) meteorological data (e.g. net radiation and precipitation).

Application of the models in BRIDGE case studies

As an illustration of how urban water balance models can be used to consider sustainable planning strategies, the application is presented of two models of the BRIDGE project, i.e. a surface atmosphere transfer scheme (SUEWS) and a hydrological model (SIMGRO). Here SUEWS is applied to Helsinki to evaluate its performance and both models are applied to the London case study. In the next sections a short description of the two sites, model set-up and scenarios are given. A more detailed description of both sites, as well as of the measurements taken at these sites is given in Chapter 5, and the urban Planning Alternatives (PA) that were considered in BRIDGE are presented in detail in Chapter 3.

Helsinki

In Helsinki, runoff observations were undertaken in three catchments (see Chapter 5, and Table 10.1): Pasila (Pa) with few pervious surfaces and a complex building and pedestrian road structure, Pihlajamäki (Pi) with much less built-up, or impervious surfaces and Veräjämäki (not included in Table 10.1) where the surface is mainly natural vegetation. The high fraction of impervious surfaces has a clear influence on the water balance. For example, in November 2010 runoff from the high-intensity built catchment was seven times higher than from the low-intensity built catchment. In November 2010, monthly precipitation of 62 mm was observed at the Kumpula site and in the low-intensity catchment 8 per

TABLE 10.1 Existing land-use cover at the Central Activity Zone in London, UK (based on the Mastermap TM, Greater London Authority of 2009) and at the two catchments Pasila (Pa) and Pihlajamäki (Pi) in Helsinki, Finland

	London Central Activity Zone	Helsinki Pa	Helsinki Pi
Buildings (-)	0.33	0.20	0.12
Streets (-)	0.34	0.42	0.22
Freshwater (-)	0.06	0	0
Grassland (-)	0.13	0.14	0.26
Coniferous trees/shrubs (-)	0	0	0.10
Broadleaf trees/shrubs (-)	0.14	0.24	0.30
Population density (# ha^{-1})	96	42	55
Surface area (ha)	2916	23.8	44.8

cent of this left the catchment as runoff, whereas in the medium- and high-intensity catchments over 50 per cent of the precipitation ended up in runoff. This shows how increasing impervious surface cover enhances the risk for urban floods, while water storage and evaporation of the vegetation cover decreases.

Here SUEWS is used to simulate summer water balance for January–August 2010 at the Pa and Pi catchments (Table 10.1). The first four months is the spin-up time to ensure such things as soil moisture and surface states are appropriate. The forcing data used were from the nearby Station for Measuring Ecosystems – Atmosphere Relations (SMEAR III) measurement station (Järvi et al. 2009). Thus the amount of precipitation at both sites is assumed to be the same (168 mm total for the period), but in Pi the amount of external irrigation is clearly smaller, causing precipitation to be the more important source of water. In Pi there is lots of natural state vegetation whereas in Pa, despite the smaller fraction of vegetation, it is more managed. At both sites, the change in soil water storage helps to maintain evaporation in summer time June to August, i.e. 6 mm and 3 mm in Pa and Pi, respectively (Figure 10.2). In the highly built-up catchment Pa, clearly less water is evaporated (111 mm y^{-1}) than in Pi (125 mm y^{-1}), and more water goes directly to runoff. SUEWS has been found to simulate the observed runoff at both catchments well (Järvi et al. 2014) and therefore can be used to examine the effect of different land uses on the water balance components.

London

In London the focus was on the Central Activity Zone (CAZ), as described in Chapter 3. Within this area there are major parks and the River Thames. The River Thames is tidal with an approximately 6 m bi-daily variation in water level in the study area. The elevation of the study area varies between 25 m above sea level in the north to sea level. The CAZ surface is highly sealed, resulting in high rainfall runoff peaks, high temperatures associated with the urban heat island and water scarcity for vegetation if no irrigation is provided. The Greater London Authority has published Planning Alternatives (PAs) to both cool the city and reduce the risk of flooding (Greater London Authority 2011a). Increasing the number of trees within the CAZ and increasing areas of green roofs are being considered to mitigate urban warming and peak runoff. The CAZ is targeted and delimited by the London planning strategies for circa 10 years and is the subject of specific planning objectives and goals, which are related to the urban metabolism (see e.g. Greater London Authority 2011a, 2011b).

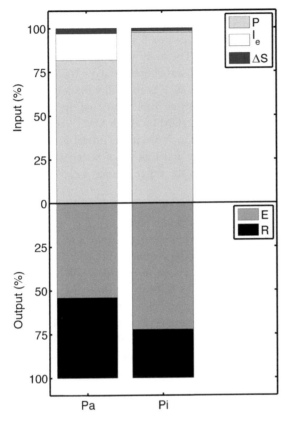

FIGURE 10.2 Simulated, using SUEWS, water balance at two catchments Pa and Pi in Helsinki, June–August 2010 P = Precipitation, Ie = irrigation, E = evaporation, ΔS = storage water change and R = surface runoff.

For London, the urban PA explored in this chapter with respect to the impact on urban water are:

- PA0: Baseline scenario: The land use map of 2009 (Mastermap TM, Greater London Authority) is used – approximately 70 per cent of the total surface being sealed and approximately 30 per cent being vegetated areas.
- PA1: Add street trees: Increase the area covered by street trees to 10 per cent of the total CAZ area (representing a 40 per cent increase of trees). The location of the trees is based on discussions with the Greater London Authority (mainly planned around sealed surfaces). The trees are assumed to be all deciduous broadleaved and the soil surface to change from impermeable to highly permeable.
- PA2: Make the roads pervious: We assumed 40 per cent of the roads evenly spread over the model grid become permeable, which is approximately 10 per cent of the total area.
- PA3: Add green roofs: The location of green roofs are based on discussions with the Greater London Authority. For a green roof to be added, the roof has to be flat or below 20 per cent and the roof area has to be larger than 25 m². In this scenario, 2 per cent of roofs in the CAZ were changed to green areas.
- PA4: Make roads pervious and add street trees: This alternative combines A1 and A2.
- PA5: Add street trees and increase the surface of green roofs: This alternative combines A1 and A3.

Simulation results for urban planning alternatives

For London SIMGRO and SUEWS were applied for the year 2010 to a 5.4 km × 5.4 km area at a resolution of 200 m × 200 m and 1000 m × 1000 m, respectively. The study area does not correspond to any water management unit or hydrological sub-catchment, but was chosen to allow the output to be combined with other models that are based on rectangular grids, such as air quality models.

To perform the SIMGRO simulations, current CAZ land cover (see Table 10.1) and the proposed PAs 1, 2 and 4 were mapped as ten Urban Morphological Types (UMTs). Each UMT represents a specific land cover type with specified reservoir characteristics. Not included in the model as presented in this paper are anthropogenic factors, such as watering gardens and leakage of water supply and sewerage pipes. In London leakage of pipes is a significant term in the water balance, i.e. 675 Ml d^{-1} leakage in 2010/2011 (Thames Water 2013).

The greener city planning scenarios are compared to a baseline (PA0: CAZ land cover in 2009) with attention to evapotranspiration, soil water shortages and intensity of runoff peaks.

The SIMGRO results show that if the number of trees or other green areas are increased, the natural soil water storage for the simulated year of 2010 will limit the evapotranspiration rate during a part of the year. If insufficient soil water is available during hot periods, evaporation will be limited and as a result the potential cooling effect by urban green will not be realised. To maintain maximum cooling, either additional irrigation water, or a greater soil water storage capacity to retain precipitation is needed. Figure 10.3 shows the relation between the fraction of impervious surface and the water limited evaporation rate for the different grid cells in the study area using the SUEWS model.

The SUEWS simulations predict that covering an additional 10 per cent of the total area with street trees (PA1) will reduce runoff by 7 per cent. In contrast, if 2 per cent of the total area is transformed

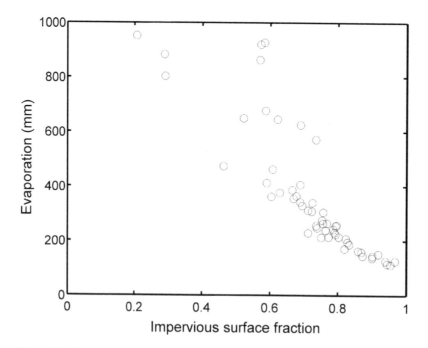

FIGURE 10.3   The relation between the fraction of impervious surface and annual evaporation rates simulated using SUEWS for individual grid cells (1 km × 1 km) of the CAZ area in London, UK

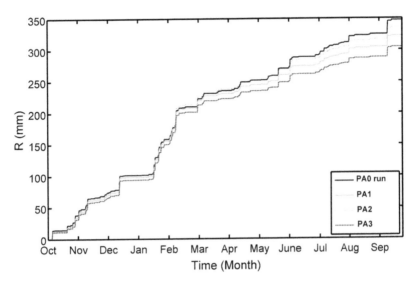

FIGURE 10.4 Cumulative hourly runoff for a highly built-up grid cell (1 km × 1 km) in central London with the different PAs, using the SUEWS model: Bas line = control (PA0), Alt1 = 40% increase of trees (PA1), Alt2 = 2% of roofs greened (PA3) and Alt3 = combined trees and green roofs (PA3)

to areas with green roofs (PA3), the reduction is 3 per cent (Figure 10.4). However, if, they are combined (PA5) there is a greater reduction in surface runoff (12 per cent) as more water is horizontally redistributed and subsequently consumed by evaporation. SIMGRO simulates very similar reductions in runoff (7 per cent) for the PA with additional street trees (PA1), while the increase in permeable streets (PA2) caused a decrease of 14 per cent in runoff, with the combination of both measures (PA4) resulting in a decrease of 17 per cent compared to the total runoff of the baseline situation (PA0).

The effects of such measures are dependent on time of year and influence one another. This is confirmed by the SIMGRO calculated excess precipitation (Figure 10.5) which indicates when additional water would be needed to sustain maximum (energy limited) evapotranspiration. In the summer months, water shortages resulting from the combined alternative (PA4) becomes larger than the permeable street (PA2) alternative. As noted above, the addition of trees in both PA1 and PA4 requires additional water to maximize evapotranspiration.

To support decision making in planning processes, spatial representation of areas at risk, or not complying to a certain threshold value represented by either a physical variable or an index, are desirable. Plate 11 presents, as an example, the areas at risk for reduced evapotranspiration rates in summer. These areas are expected to be associated with warmer temperatures because of reduced cooling. For the combined alternative (PA4) there is a decrease of 25 per cent in risk of warmer temperatures because of lack of cooling compared to the baseline (PA0); for infiltrating streets (PA2) a decrease of 20 per cent; and for trees (PA1) 15 per cent. The areas of medium risk increase for every alternative: for the combination (PA4) with 15 per cent, for streets (PA2) 10 per cent, and for trees (PA1) 5 per cent. These changes in risk levels can be directly related to the land cover changes of the alternatives. These simulations show (see Plate 11) that increase of urban green space in combination with water availability (necessary to support the evaporation rates) does increase actual evapotranspiration which contributes to a decrease of city temperatures.

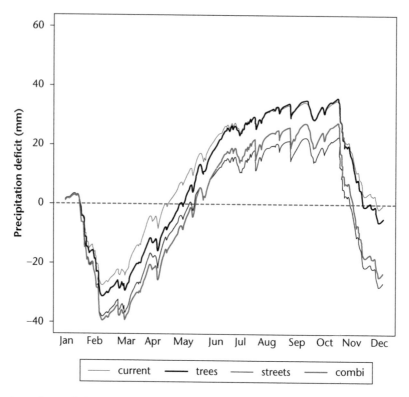

FIGURE 10.5 Annual cumulative precipitation shortage (negative numbers indicate precipitation excess) in 2010 for the control and the three PAs, using the SIMGRO model for the CAZ area in London, UK: current = 70% sealed and 30% of area green (PA0), trees = 40% increase of trees (PA1), streets = 40% more permeable streets (PA2) and combi = combined (PA4)

Discussion and conclusions

Risk maps give a quick overview of the effects of the implementation of urban PAs and can be useful for an evaluation and understanding the consequences. In this study, alternative strategies for a greener city, the CAZ of London, UK, were compared with a baseline and evaluated focussing on the effects of an increase of actual evapotranspiration and a decrease of intensity in rainfall generated runoff peaks. The risk maps (see Plate 11) showed that additional urban green space is most effective in dominantly sealed areas.

If the goal is to use evaporation increases to lower air temperature for human comfort, not only the evaporation rate but also the meteorological conditions will be important (e.g. Müller et al. 2013). Obviously, the greatest opportunity for change is in the areas with a large fraction of sealed surfaces. However, to make such a shift possible in these areas will mean a drastic land use change from built-up to vegetated surfaces, or open water. To implement such alternatives sufficient water needs to be available at the appropriate time. This may require irrigation.

Not surprisingly, the quality of the results is highly dependent on the quality of the input data (forcing data, model parameters) and models used. It is essential that the models are extensively evaluated to ensure that they are capable of performing appropriately for both current conditions and proposed PAs. Thus the measurements of evaporation (see Chapter 5 for details) and evaluation of these processes (e.g. Chapter 9) are critical to not only energy but also to water exchange related processes. Models, such as presented in this chapter, allow for a rapid assessment of different PAs, while simultaneously providing insights in the dynamic processes and interactions between different components of the hydrological cycle in the urban environment. As such these models can aid in the identification of potential problem areas associated with different urban PAs and support planning processes at the relatively large scale of settlements and master plans.

References

Berthier, E., Andrieu, H. & Creutin, J. D. (2004). The role of soil in the generation of urban runoff: development and evaluation of a 2D model. *Journal of Hydrology*, 299, 252–266.

Berthier, E., Dupont, S., Mestayer, P. G. & Andrieu, H. (2006). Comparison of two evapotranspiration schemes on a sub-urban site. *Journal of Hydrology*, 328, 635–646.

Chrysoulakis, N., Lopes, M., San José, R., Grimmond, C. S. B., Jones, M.B ., Magliulo, V., Klostermann, J. E. M., Synnefa, A., Mitraka, Z., Castro, E., González, A., Vogt, R., Vesala, T., Spano, D., Pigeon, G., Freer-Smith, P., Staszewski, T., Hodges, N., Mills, G., & Cartalis, C. (2013). Sustainable urban metabolism as a link between bio-physical sciences and urban planning: the BRIDGE project. *Landscape and Urban Planning*, 112, 100–117.

Diaper, C. & Mitchell, V. G. (2007). Urban Volume and Quality (UVQ). *Urban Water Resources Toolbox: Integrating Groundwater Resources into Urban Water Management*, Wolf, L., Morris, B., and Burn, S., Ed., IWA Publishing, London, 297 pp.

Dupont, S., Mestayer, P. G., Guilloteau, E., Berthier, E. & Andrieu, H. (2006). Parameterization of the urban water budget with the Submesoscale Soil Model. *Journal of Applied Meteorology and Climatology*, 45, 624–648.

Feddes, R. A., Kabat, P., van Bakel, P. J. T., Bronswijk, J. J. B. & Halbertsma, J. (1988). Modeling soil water dynamics in the unsaturated zone – state of the art. *Journal of Hydrology*, 100, 69–111.

Greater London Authority (2011a). Securing London's water future. The Mayor's Water Strategy. City Hall, The Queen's Walk, London, UK, 116 pp.

Greater London Authority (2011b). Managing risks and increasing resilience. The Mayor's Climate Change Adaptation Strategy. City Hall, The Queen's Walk, London, UK, 126 pp.

Grimmond, C. S. B., Oke, T. R. & Steyn, D. G. (1986). Urban water balance: 1. A model for daily totals. *Water Resources Research*, 22, 1397–1403.

Grimmond, C. S. B. & Oke, T. R. (1986). Urban Water-Balance 2: Results from a suburb of Vancouver, British-Columbia. *Water Resources Research*, 22, 1404–1412.

Grimmond, C. S. B. & Oke, T. R. (1991). An evapotranspiration-interception model for urban areas. *Water Resources Research*, 27, 1739–1755.

Järvi, L., Hannuniemi, H., Hussein, T., Junninen, H., Aalto, P. P., Hillamo, R., Makela, T.,–Keronen, P., Siivola, E., Vesala, T. & Kulmala, M. (2009). The Urban Measurement Station Smear III: Continuous monitoring of air pollution and surface-atmosphere interactions in Helsinki, Finland, *Boreal Environment Research*, 14, 86–109.

Järvi, L., Grimmond, C. S. B. & Christen, A. (2011). The Surface Urban Energy and Water Balance Scheme (SUEWS): Evaluation in Los Angeles and Vancouver. *Journal of Hydrology*, 411, 219–237.

Järvi, L., Grimmond, C. S. B., Taka, M., Nordbo, A., Setälä, H. & Strachan, I. (2014) Developments of the Surface Urban Energy and Water balance Scheme (SUEWS) for cold climate cities. *Geoscientific Model Developments Discussion*, 7, 1063–1114.

Jia, Y. W., Ni, G. H., Kawahara, Y. & Suetsugi, T. (2001). Development of WEP model and its application to an urban watershed. *Hydrological Processes*, 15, 2175–2194.

Lemonsu, A., Masson, V. & Berthier, E. (2007). Improvement of the hydrological component of an urban soil-vegetation-atmosphere-transfer model. *Hydrological Processes*, 21, 2100–2111.

Masson V. (2000). A physically-based scheme for the urban energy budget in atmospheric models. *Boundary-Layer Meteorology*, 94, 357–397.

Mitchell, V. G., Mein, R. G. & McMahon, T. A. (2001). Modelling the urban water cycle. *Environmental Modelling & Software*, 16, 615–629.

Mitchell, V. G., McMahon, T. A. & Mein, R. G. (2003) Components of the total water balance of an urban catchment. *Environmental Management*, 32, 735–746.

Mitchell, V. G. & Diaper, C. (2005). UVQ: A tool for assessing the water and contaminant balance impacts of urban development scenarios. *Water Science and Technology*, 52(12), 91–98.

Müller, N., Kuttler, W. and Barlag, A.-B. (2013). Counteracting urban climate change: adaptation measures and their effect on thermal comfort. *Theoretical and Applied Climatology*, DOI 10.1007/s00704–013–0890–4.

Noilhan, J. & Planton, S. (1989). A simple parameterization of land surface processes for meteorological models. *Monthly Weather Review*, 117, 536–549.

Rodriguez, F., Andrieu, H. & Zech, Y. (2000). Evaluation of a distributed model for urban catchments using a 7-year continuous data series. *Hydrological Processes*, 14, 899–914.

Rodriguez, F., Andrieu, H. & Morena, F. (2008). A distributed hydrological model for urbanized areas; Model development and application to case studies. *Journal of Hydrology*, 351, 268–287.

Thames Water (2013). Final Drought Plan. Main Report. Thames Water Utilities Limited, Clearwater Court, Vastern Road, Reading, UK. 151 pp.

van Walsum, P. E. V. & Veldhuizen, A. A. (2011). Integration of models using shared state variables: Implementation in the regional hydrologic modelling system SIMGRO, *Journal of Hydrology*, 409, 363–370.

Wang, J., Endreny, T. A. & Nowak, D. J. (2008). Mechanistic simulation of tree effects in an urban water balance model. *Journal of the American Water Resources Association*, 44, 75–85.

Wolf, L., Morris, B. & Burn, S. (2007). *Urban Water Resources Toolbox: Integrating Groundwater into Urban Water Management*. London, IWA, 309 pp.

11

URBAN CARBON BUDGET MODELLING

Donatella Spano[1,2], Serena Marras[1,2] and Veronica Bellucco[1]

[1]UNIVERSITY OF SASSARI AND [2]EURO-MEDITERRANEAN CENTRE ON CLIMATE CHANGE

Introduction

It is well known that carbon dioxide (CO_2) concentration in the atmosphere increased from 278 parts per million (ppm), at the beginning of the Industrial Era (1750), to 391.4 ppm at the end of 2011 (Conway & Tans 2012). Anthropogenic activities, such as deforestation and changes in land use, were the dominant source of CO_2 emissions until the Industrial Revolution (around 1920), when the combustion of fossil fuel became dominant. From 2002 to 2011, fossil fuel emissions contributed 89 per cent to the total emissions at global level, while changes in land use accounted for the 11 per cent (Le Quéré et al. 2013). Projections for 2012 reveal that global emissions will reach 9.7 ± 0.5 petagrams of carbon (PgC), which is 2.6 per cent above the 2011 levels.

Urban areas are recognized to constitute the major source of the CO_2 emitted into the atmosphere. Svirejeva-Hopkins et al. (2004) suggest that more than 90 per cent of anthropogenic carbon emissions are generated in cities, and human activities (human respiration, domestic heating/cooling, transportation, etc.) produce more than 70 per cent input of CO_2 into the urban environment (Canadell et al. 2009).

Thus, cities change the carbon cycle, the atmosphere and climate (Mills 2007), but climate itself influences urban matter and energy fluxes. The link between urbanization and global climate change is complex, and the knowledge of urban carbon exchanges is important to better understand the interaction of natural and anthropogenic processes that control the role of cities in the global warming phenomenon (Walsh et al. 2004). Also, a crucial point in urban sustainable development is to evaluate the impact that future urban planning alternatives may have on the main factors affecting the citizens' well-being and health, especially related to increases in carbon emissions.

As part of the BRIDGE project (Chrysoulakis et al. 2013), carbon exchanges have been measured (see Chapter 5 for details) and simulated. Models' results are reported in this chapter. First, a description of the main emission sources in a city is presented.

Urban carbon budget

The urban metabolism concept has been applied to energy and water, but the carbon balance can also be addressed as an urban metabolic process. This approach is used to simplify the complexity of inputs,

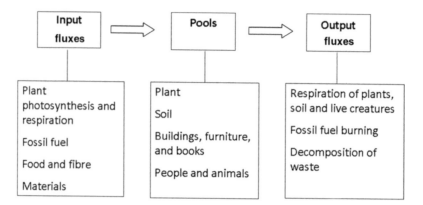

FIGURE 11.1 Carbon input, pools and output in urban ecosystems (modified from Churkina 2008)

outputs and interactions of carbon exchanges in an urban system. A conceptual scheme of the urban carbon budget is shown in Figure 11.1 (Churkina 2008). Natural and anthropogenic inputs contribute to the carbon storage in different pools (e.g. vegetation, soil, building and human products, people, and animals). Urban carbon outputs include respiration by live creatures and soils, burning of fossil fuel, and waste decomposition. All outputs are quantified as CO_2 emissions.

The form and structure of cities, in addition to building characteristics, may well influence the amount of CO_2 emitted and stored. In addition, demographic trends affect the urban forms in ways that have an impact on the CO_2 emissions. The pattern of urban development may be the key to determining the amount of CO_2 emissions and the ability of cities to reduce them (Alberti 2008). The number and size of households affect the number and size of buildings and associated energy uses. Furthermore, the spatial distribution of residential and commercial housing units affects commuting patterns and transportation choices with important consequences for fossil fuel consumption. Planning strategies should then consider these aspects to reduce carbon emissions and to increase carbon storage in human settlements.

Modelling urban carbon flux

Modelling urban CO_2 fluxes is challenging, especially at local scale, because it must account for different characteristics of a specific urban area. Two approaches are often used or combined to model net CO_2 emissions:

- a top-down approach, which elaborates and separates information from a higher to a lower scale;
- a bottom-up approach, which analyses information and generalizes emissions from a local to a higher scale.

Atmospheric CO_2 emissions together with uptake from urban vegetation represent the net exchange of vertical fluxes between urban surfaces and the upper atmosphere. Several studies (Nemitz et al. 2002; Moriwaki & Kanda 2004; Crawford et al. 2011; Christen et al. 2011) have defined the net carbon dioxide flux (Fc) as the sum of different sources, such as traffic (E_{Tr}) and building (E_B) emissions, human and animal respiration (R_A), below-ground autotrophic and heterotrophic respiration (R_S), and above-ground vegetation respiration minus the CO_2 uptake by photosynthesis ($R_V - P_V$) (modified from Crawford et al. 2011):

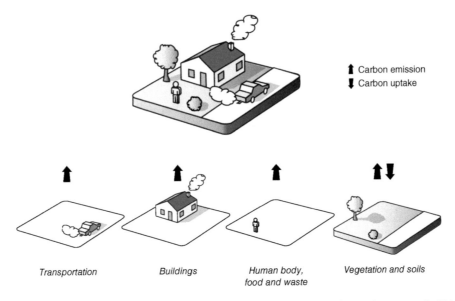

Carbon emission
Carbon uptake

Transportation   Buildings   Human body,   Vegetation and soils
                                            food and waste

FIGURE 11.2   Source of CO_2 emissions and uptake in urban areas (modified from Christen et al. 2010)

$$Fc = E_{Tr} + E_B + R_A + R_S + (R_V - P_V)$$

The emission sources are derived from sub-models (Figure 11.2) that require information on urban structure, buildings and land cover, together with meteorological, vehicles, census, and statistical data of population, land use and employment. On average, annual emissions vary from 9.2 μmol m^{-2} s^{-1} in the city centre of Copenhagen (Soegaard & Møller-Jensen 2003) to 14.7 μmol m^{-2} s^{-1} for the urban site in Montreal, and 15.2 μmol m^{-2} s^{-1} in the suburban neighbourhood of Vancouver (Christen et al. 2011). Differences between modelled and measured fluxes are generally in order of 10 per cent of the mean flux density. A brief description on how to model each CO_2 emission source is reported below.

Vehicles emissions (E_{Tr})

Traffic is the most important source of CO_2 fluxes in urban areas, with a typical bimodal daily profile due to the rush hours on weekdays (Figure 11.3) (Bergeron & Strachan 2011). The number of vehicles is almost constant throughout the year, with a reduction during summer holidays. For example, a reduction of about 20 per cent was observed in Copenhagen by Soegaard & Møller-Jensen (2003). A strong linear relationship between Fc and vehicular traffic emissions was found for the urban areas of Edinburgh, Copenhagen and Montreal, and the difference in intercepts between summer and winter season (Figure 11.3) confirmed the presence of other non-traffic sources (i.e. winter local heating) (Nemitz et al. 2002; Soegaard & Møller-Jensen 2003; Bergeron & Strachan 2011). So, traffic is not the major contributor to seasonal variations of measured vertical CO_2 fluxes (Moriwaki & Kanda 2004).

Different top-down approaches have been used to estimate vehicular traffic emissions. They have been calculated as residual between the total CO_2 flux, measured by Eddy Covariance (EC) and the other summed estimated emission sources (Nemitz et al. 2002). When local traffic counts datasets are available, E_{Tr} is estimated by multiplying the total consumption of fuel by emission coefficients (Moriwaki & Kanda 2004). It is important to know the vehicle type, the travelled distance and the fuel

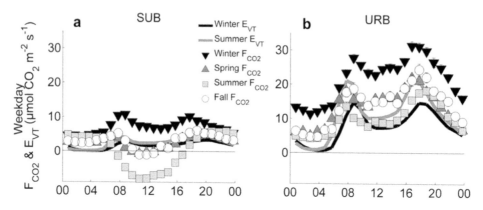

FIGURE 11.3 Weekday summer and winter profiles of vehicular CO_2 emissions (E_{VT}) compared to total CO_2 emissions (F_{CO2}) for an urban (URB) and suburban (SUB) area of Montreal (modified from Bergeron & Strachan 2011)

type (Bergeron & Strachan 2011; Christen et al. 2011) to have more accurate estimates. For example, following Bergeron and Strachan (2011), E_{Tr} is calculated by multiplying the per vehicle daily vehicle distance ($pvDVD$, km vehicle^{-1} day^{-1}) by the vehicles owned by each person (NV, vehicle person^{-1}), the fraction of daily traffic per hour (Ft, day hour^{-1}), the population density (ρ_{pop}, person m^{-2}), and by an emission coefficient (EFC, µmol CO_2km^{-1}):

$$E_{TR} = pvDVD \cdot NV \cdot Ft \cdot \rho_{pop} \cdot EFC$$

Estimated values of total annual carbon and fuel consumption for each type of vehicle and fuel are shown in Table 11.1 (Christen et al. 2010). Daily traffic emissions values vary from 11.7 µmol m^{-2} s^{-1} in Edinburgh (Nemitz et al. 2002), to 7.5 µmol m^{-2} s^{-1} in Tokyo (Moriwaki & Kanda 2004), whereas an annual value of 7.7 µmol m^{-2} s^{-1} was found in Vancouver (Christen et al. 2011). It is estimated that E_{Tr} contributes approximately 40–50 per cent of total Fc (Table 11.2).

Emissions from buildings (E_B)

Emissions resulting from heating buildings and other household services represent the second most important CO_2 source. Emission estimates from buildings is based on local fossil fuel consumption

TABLE 11.1 Carbon emissions factors and fuel consumption per vehicle and fuel type (Christen et al. 2010). Vehicles are divided in four categories based on their task and travelled distance

Vehicle type	Fuel	Fuel efficiency	
		(l km^{-1})	(g C km^{-1})
Light	Gasoline	0.115	75.6
Medium freight	Gasoline	0.265	174.3
Heavy freight	Diesel	0.39	296.0
Transit	Diesel	0.39	296.0

TABLE 11.2 Modelled annual CO_2 fluxes, separated by source, in the suburban neighbourhood of Vancouver (modified from Christen et al. 2010). Units are expressed in $\mu mol\ m^{-2}\ s^{-1}$ and percentage of contribution to the total flux

				Vertical flux (F_C) 15.16			
Buildings (E_B) 6.52 (40%)		Transportation (E_{Tr}) 7.74 (47%)		Human respiration (R_A) (8%)	Vegetation −0.42 (14%)		
Residential	Other	Local traffic	Through traffic		Soil respiration (R_S)	Above-ground respiration (R_V)	Photo-synthesis (P_V)
5.68	0.84	0.82	6.92	1.29	0.74	0.13	−1.29

statistics (Nemitz et al. 2002), or energy use data multiplied by the density of dwellings (Moriwaki & Kanda 2004). Recently, a more detailed bottom-up building typology approach has been developed for the city of Vancouver. Buildings have been classified based on their use, form, date of construction and energy consumption by using remote sensing (LiDAR) and statistical data (Christen et al. 2010, 2011). Two building energy models, based on building volume and typology attributes, have been used to estimate E_B emissions separately for residential dwellings (95 per cent of total buildings) and "other" buildings (5 per cent of total buildings).

Buildings' contribution from natural gas combustion varies from 12.6 $\mu mol\ m^{-2}\ s^{-1}$ per day (52 per cent of total net CO_2 flux) in Edinburgh (Nemitz et al. 2002) to seasonal values of 8.04 $\mu mol\ m^{-2}\ s^{-1}$ in winter and 2.23 $\mu mol\ m^{-2}\ s^{-1}$ in summer in Tokyo (Moriwaki & Kanda 2004). An annual value of 6.52 $\mu mol\ m^{-2}\ s^{-1}$ (40 per cent of total) was found by Christen et al. (2011) in Vancouver (Table 11.2).

Human respiration (R_A)

Human respiration can be considered the third most important source to Fc. The contribution of human respiration to Fc is calculated by multiplying the estimated CO_2 human emission by population density (assuming homogeneous distribution of population over the study area or estimated based on residential floor area by using LiDAR volume and land use data). Daily human exhalation varies from 201.5 $\mu mol\ CO_2\ person^{-1}\ s^{-1}$ (Moriwaki & Kanda 2004) to 280 $\mu mol\ CO_2\ person^{-1}\ s^{-1}$ (Nemitz et al. 2002).

Mean daily R_A values of about 2 $\mu mol\ CO_2\ m^{-2}\ s^{-1}$ were found in Edinburgh (Nemitz et al. 2002) and Tokyo (Moriwaki & Kanda 2004), while an annual contribution of 1.3 $\mu mol\ CO_2\ m^{-2}\ s^{-1}$ was found in the suburban area of Vancouver (Christen et al. 2011) (Table 11.2).

Soil and vegetation emissions and uptake

Daytime uptake for vegetation photosynthesis could reduce the net emissions generated from other CO_2 sources. Suburban areas, with more vegetation than city centres, show a similar behaviour to natural ecosystems (Crawford et al. 2011). As vegetation cover decreases, Fc increases because of less plant uptake (Figure 11.4) (Velasco & Roth 2010; Bergeron & Strachan 2011). This effect is particularly evident during the plants' growing season. In cities with less than 34 per cent of vegetation cover, plants are not more able to offset other sources of CO_2 emissions (Bergeron & Strachan 2011). Emissions and uptake from soil and vegetation can be modelled separately using the above equation for the total carbon dioxide flux estimation.

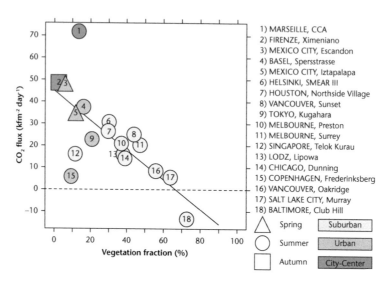

FIGURE 11.4 F_C as function of different vegetation cover fraction (from Velasco & Roth 2010)

The CO_2 emission due to respiration of non-sealed surfaces (R_S) is considered negligible when their extension represents only a small percentage of the total area (Moriwaki & Kanda 2004). According to Lloyd & Taylor (1994) and Liss et al. (2009), R_S is a function of soil temperature (Soegaard & Møller-Jensen 2003) and volumetric water content (Christen et al. 2011). To up-scale the flux estimate, the calculated R_S is multiplied by the non-sealed area fraction (Soegaard & Møller-Jensen 2003; Christen et al. 2011).

The CO_2 emissions from vegetation respiration (R_V) can be estimated from measurements of dark respiration in leaves or needles of representative tree species in the study area, or calculated as a function of air temperature. Vegetation cover fraction or, recently, satellite-estimated Leaf Area Index (LAI), as well as climate data, are used to up-scale R_V (Nemitz et al. 2002; Christen et al. 2011).

Uptake from photosynthesis (P_V) can be modelled in different ways: (i) as function of light response curves (Nemitz et al. 2002; Bergeron & Strachan 2011; Christen et al. 2011); (ii) using meteorological data and satellite LAI estimated as input of a "leaf level photosynthesis" model (Soegaard & Møller-Jensen 2003); and (iii) from literature data for similar vegetation type and scaling up the flux depending on the vegetation cover fraction (Moriwaki & Kanda 2004).

Christen et al. (2011) reported annual values for each of the three components (Table 11.2). R_S, R_V and P_V represent a total net contribution to CO_2 flux of -0.42 μmol m^{-2} s^{-1}. Vertical carbon fluxes from vegetation and soil have been also estimated by using ecosystem process models. They can simulate responses of vegetation to urban pollution, enhanced levels of atmospheric CO_2 and to changes in urban climate (such as the BIOME BioGeochemical Cycles (BIOME-BGC) model, and the Carnegie-Ames-Stanford-Approach (CASA) model). The effects of land use changes on carbon balance have been investigated through statistical models driven by changes in urban population (Svirejeva-Hopkins et al. 2004).

All models, however, only focused on one effect (e.g. climate change, vegetation or land cover conversion), but to have a good assessment of an urban system's impacts on the global carbon cycle, models should include both biophysical and human dimensions and their interactions (Figure 11.2).

Application and results

In the BRIDGE project, the Soil-Vegetation-Atmosphere-Transport (SVAT) model Advanced Canopy-Atmosphere-Soil Algorithm (ACASA) was improved to identify how multiple environmental factors, such as climate variability, population density and species distribution, impact future carbon cycle prediction in urban environment. A short description of the model has been given in Chapter 9. A brief description on how the model estimates CO_2 flux is reported below. The model was used at both local and regional scales, when nested, as a land surface scheme, in the meso-scale Weather Research and Forecasting (WRF) model.

ACASA model description

The total carbon dioxide flux in ACASA is estimated considering both the human emissions and emissions and uptake by vegetation and soil (Falk et al. in review). Biotic carbon assimilation is estimated using the equations from Farquhar and von Caemmerer (1982) and Harley et al. (1992). The dark respiration rates of CO_2 exchange between the leaf and the atmosphere are calculated separately using an Arrhenius-type, temperature-based exponential Q_{10} relationship for each micro-scale surface type and situation.

As a default, vegetation density is assumed to decrease linearly with population density above 1000 people km^{-2}, with zero values for population densities exceeding 10,000 people km^{-2}. The mean *LAI* is set to 0.5 for cities of moderate to high urban development.

Anthropogenic sources considered by the model include fuel combustion due to transportation (E_{Tr}) and human bodily respiration components (R_A). Human bodily respiration is set to ~0.14 g CO_2 person^{-1} minute^{-1}. When available, measured hourly traffic data converted to number of vehicles per m^{-2} (*Tr*) as well as the length of the road (*lr, km*), and an averaged emission factor for the vehicles travelling on the road (*EFC*, μmol CO_2 vehicle^{-1} km^{-1}) per time step $(\Delta t, s)$ are used to estimate the carbon emissions from transportation as:

$$E_{Tr} = EFC \cdot lr \cdot Tr \cdot \Delta t^{-1}$$

Next steps in ACASA development will be to include the energy buildings contribute to have the complete estimation of the all anthropogenic components contributing to the total CO_2 flux over cities.

ACASA model application

In the BRIDGE project, the ACASA model was applied both at local and regional scales to estimate carbon exchanges over Helsinki and Firenze cities. *In situ* EC measurements were used to calibrate and validate the model at the local scale, whereas the coupled model WRF-ACASA was run by using five nested domains at 48.6 km × 48.6 km (1); 16.2 km × 16.2 km (2); 5.4 km × 5.4 km (3); 1.8 km × 1.8 km (4); and 0.6 km × 0.6 km (5) horizontal grid space, respectively. The ratio between each parent and nested domain was 1:3.

In addition, the model was used in combination with a Cellular Automata (White et al. 1997; White & Engelen 2000) and a transportation model (Tsekeris & Stathopoulos 2006, 2009) to predict the impact of future planning choices in the Firenze case study. A brief description of the main results is reported here.

Helsinki

Data collected at the Station for Measuring Ecosystems – Atmosphere Relations (SMEAR) III measurement station (60°12′N, 24°57′E, 21 m above sea level), approximately four kilometres northeast

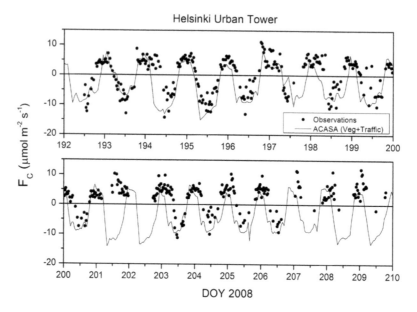

FIGURE 11.5   Time series plot showing measured (flux tower) and modelled (WRF_ACASA tower pixel) carbon flux for Helsinki for the period 19–29 July 2008, modified from Falk et al. (in review)

from the city centre, were used to validate model results in the tower pixel. Vegetation cover percentage surrounding the tower range from 41 per cent to 50 per cent with vegetation being the dominant land cover type in all three sectors described by Järvi et al. (2012) as vegetated, road and built-up. Following Järvi et al. (2012), a typical traffic distribution of 83 per cent passenger cars, 10 per cent vans, 3 per cent trucks and 4 per cent buses (Lilleberg & Hellman 2011) was assumed, giving a mixed fleet CO_2 emission factor of 285 g km^{-1}. Results for some days in July 2008 are shown in Figure 11.5. Fc shows significant uptake during the daytime both in the observations and the modelled time series, indicating significant vegetation contribution in the flux footprint. A clear diurnal pattern with a major rush hour traffic peak in the afternoon and a smaller one in the morning can be observed, as well as a decreased Fc during the weekends. At the regional scale, results for domain 5 are showed in Plate 12. One of the strengths of the WRF-ACASA model is that it reproduces realistic maps of the urban fluxes for current conditions, but it is also able to produce estimates for future conditions, as described for the Firenze case study below.

Firenze

Besides local and regional estimates of CO_2 fluxes in the Firenze city centre (43°47′ N, 11°15′ E) (not shown), ACASA model was used to estimate the impact of changes in land use, by integrating the coupled version WRF-ACASA with a Cellular Automata (CA), which simulated the urban land-use dynamics, and a Transportation model, which estimated the variation of the transportation network load (Figure 11.6; Blecic et al. 2011). The CA module generates future land-use scenarios, as maps, (Plate 13) by using information on the current land uses and street network, the zoning regulations, the physical suitability of the cells to develop into specific land uses, and a set of alternative projections of the aggregate demand for different land uses. In a second step, the maps of future land uses, together with the street network, are fed into the transportation module to estimate future road traffic. The

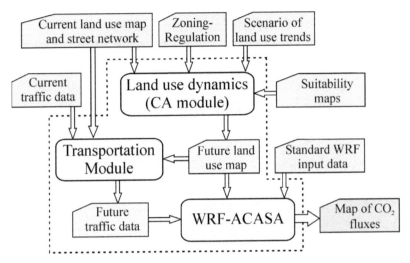

FIGURE 11.6 Modelling framework with the most relevant data exchanges between models (modified from Blecic et al. 2011)

current road traffic data are used for its calibration. Finally, WRF-ACASA model uses both future land uses and road traffic to simulate the carbon exchanges between the urban surface and the atmosphere, and their interaction with local weather. WRF-ACASA outputs are then future maps of CO_2 fluxes in the urban area under consideration (Figure 11.7).

The 100 m-resolution CORINE Land Cover 2000 (CLC2000) dataset was used as the input land-use layer, as well as the Firenze master plan, which regulates the land-use permissions and prohibitions relevant for the CA simulation. The transportation module was designed to import spatial data from the OpenStreetMap collaborative project, which provides free street network data with a notable level of detail and geographic coverage. A pre-processing phase for the calculation of the cells' accessibility was carried out using the current street network. A future land-use projection corresponding to a 20-year evolution of the urban area, which showed a 10 per cent increase continuous urban fabric, 20 per cent increase in discontinuous urban fabric and 10 per cent increase in commercial/industrial land uses, was then produced as well as the variation of the load of vehicles on the road network (Plate 13).

Finally, the WRF-ACASA model generated maps of CO_2 fluxes by using the land-use scenario and traffic load information as input. Results for CO_2 fluxes for April 2008 for the current land uses (based on CLC2000) and the 2020 future land-use scenario are reported in Figure 11.7. The increased urbanization level causes the change of the investigated area from being a net sink to becoming a net source of CO_2, with values passing from -2 μmol m^{-2} s^{-1} to 2 μmol m^{-2} s^{-1} (Blecic et al. 2014). Green areas have become urbanized under the land-use scenario, therefore emitting rather than contributing to the absorption of the local CO_2 fluxes. In addition, the impact on the net variation of emissions is not only restricted to the new urbanized area in 2020, the increase in urbanization can also have an amplified impact on CO_2 emissions, which covers a more wide area.

Conclusions

Urban systems are extremely complex and the interaction of carbon emissions with the environment and local meteorology make it a non-trivial task to quantify and predict CO_2 fluxes in order to support urban planning and management in a more sustainable way. Different approaches and methods are used

Current land use (2000)

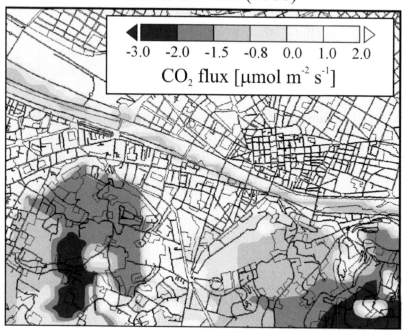

Future land use (2020)

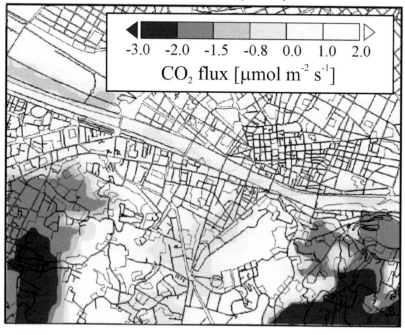

FIGURE 11.7 WRF-ACASA simulated CO_2 fluxes for current (top panel) and future 2020 (bottom panel) land-use scenario (modified from Blecic et al. 2014)

to model CO_2 flux over cities, and, in the BRIDGE project, an off-line coupling between different models was proposed to link urban planning decisions to the estimates of the environmental impact, especially in terms of CO_2 emissions in the atmosphere.

Models and methodologies to improve urban CO_2 flux estimates require further testing and development, but they represent a promising perspective for developing Decision Support Systems for sustainable urban planning.

References

Alberti, M. (2008). *Advances in Urban Ecology: Integrating Humans and Ecological Processes in Urban Ecosystems.* Springer, 366 pp.

Bergeron, O. & Strachan, I. B. (2011). CO_2 sources and sinks in urban and suburban areas of a northern mid-latitude city, *Atmospheric Environment*, 45, 1564–1573.

Blecic I., Cecchini, A., Falk, M., Marras, S., Pyles, R. D., Spano, D. & Trunfio, G. A. (2011). Towards a planning Decision Support System for low-carbon urban development. *Computational Science and Its Applications* - ICCSA 2011. ISBN 978–3–642–21927–6.

Blecic I., Cecchini, A., Falk, M., Marras, S., Pyles, R. D., Spano, D. & Trunfio, G. A. (2014). Urban metabolism and climate change: a planning support system. *International Journal of Applied Earth Observation and Geoinformation*, 26, 447–457.

Canadell J. G., Ciais, P., Dhakal, S., Le Quéré, C., Patwardhan, A. & Raupach, M. R. (2009). *The Human Perturbation of the Carbon Cycle.* Pairs, France: UNESCO-SCOPE-UNEP.

Christen, A., Coops, N., Kellett, R., Crawford, B., Heyman, E., Olchovski, I., Tooke, R. & van der Laan, M. (2010). *A LiDAR-Based Urban Metabolism Approach to Neighbourhood Scale Energy and Carbon Emissions Modelling.* University of British Columbia, Technical report prepared for Natural Resources Canada, 104 pp.

Christen A., Coops, N. C., Crawford, B. R., Kellett, R., Liss, K. N., Olchovski, I., Tooke, T. R., van der Laan, M. & Vogt, J. A. (2011). Validation of modeled carbon-dioxide emissions from an urban neighborhood with direct eddy-covariance measurements. *Atmospheric Environment*, 45, 6057–6069.

Chrysoulakis, N., Lopes, M., San José, R., Grimmond, C. S. B., Jones, M. B., Magliulo, V., Klostermann, J. E. M., Synnefa, A., Mitraka, Z., Castro, E., González, A., Vogt, R., Vesala, T., Spano, D., Pigeon, G., Freer-Smith, P., Staszewski, T., Hodges, N., Mills, G., & Cartalis, C. (2013). Sustainable urban metabolism as a link between bio-physical sciences and urban planning: the BRIDGE project. *Landscape and Urban Planning*, 112, 100–117.

Churkina, G. (2008). Modeling the carbon cycle of urban systems. *Ecological Modeling* 216, 107–113.

Conway, T. J. & Tans, P. P. (2012). *Trends in Atmospheric Carbon Dioxide*, http//www.esrl.noaa.gov/gmd/ccgg/trends (last accessed 16 November 2012).

Crawford B., Grimmond, C. S. B. & Christen, A. (2011). Five years of carbon dioxide fluxes measurements in a highly vegetated suburban area, *Atmospheric Environment*, 45, 896–905.

Falk, M., Pyles, R. D., Marras, S., Spano, D., Paw U, K. T. & Snyder, R. L. (in review). Regional simulation of urban metabolism in Helsinki, Finland in 2008.

Farquhar, G. D. & von Caemmerer, S. (1982). Modeling photosynthetic response to environmental conditions. Physiol. Plant Ecol. II. In *Encyclopedia of Plant Physiology New Series*, 12. Lange, O. L., P. S. Nobel, C. B. Osmond, H. Ziegler, Ed. Springer-Verlag, Berlin, Germany, p. 747.

Harley P. C., Thomas, R. B., Reynolds, J. F. & Strain, B. R. (1992). Modelling photosynthesis of cotton grown in elevated CO_2. *Plant Cell Environ.*, 15, 271–282.

Järvi L., Nordbo, A., Junninen, H., Riikonen, A., Moilanen, J., Nikinmaa, E. & Vesala, T. (2012). Seasonal and annual variation of carbon dioxide surface fluxes in Helsinki, Finland, in 2006–2010. *Atmos. Chem. Phys. Discuss.* 12, 8355–8396.

Le Quéré C., Andreas, R. L., Boden, T., Conway, T., Houghton, R. A., House, J. I., Marland, G., Peters, G. P., van der Werf, G., Ahlstrom, A., Andrew, R. M., Bopp, L., Canadell, J. G., Ciais, P., Doney, S. C., Enright, C., Friedlingstein, P., Huntingford, C., Jain, A. K., Jourdain, C., Kato, E., Keeling, R. F., Klein Goldewijk, K., Levis, S., Levy, P., Lomas, M., Poulter, B., Raupach, M. R., Schwinger, J., Sitch, S., Stocker, B. D., Viovy, N., Zaehle, S. & Zeng, N. (2013). The global carbon budget 1959–2011. *Earth Syst. Sci. Data*, 5, 165–185.

Lilleberg I. & Hellman, T. (2011). The development of traffic in Helsinki in 2010 in *Helsinki City Planning Department* 2011:2, http://www.hel2.fi/ksv/julkaisut/los_2011–2.pdf (in Finnish, last accessed November 2012).

Liss K., Crawford, B., Jassal, R., Siemens, C. & Christen, A. (2009). Soil respiration in suburban lawns and its response to varying management and irrigation regimes. In *Proc. of the AMS Eighth Conference on the Urban Environment*. Phoenix, AZ, 11–15 January.

Lloyd, J. & Taylor, J. A. (1994). On the temperature dependence of soil respiration. *Functional Ecology*, 8, 315–323.

Mills, G. (2007). Cities as agents of global change. *International Journal of Climatology*, 27, 1849–1857.

Moriwaki R. & Kanda, M. (2004). Seasonal and diurnal fluxes of radiation, heat, water vapour, and carbon dioxide over a suburban area. *Journal of Applied Meteorology*, 43, 1700–1710.

Nemitz E., Hargreaves, K. J., McDonald, A. G., Dorsey, J.R. & Fowler, D. (2002). Meteorological measurements of the urban heat budget and CO_2 emissions on a city scale, *Environ. Sci. Technol.*, 36, 3139–3146.

Soegaard, H. & Møller-Jensen, L. (2003). Towards a spatial CO_2 budget of a metropolitan region based on textural image classification and flux measurements, *Remote Sensing Environment*, 87, 283–294.

Svirejeva-Hopkins, A., Schellnhuber, H. J. & Pomaz, V. L. (2004). Urbanized territories as a specific component of the Global Carbon Cycle. *Ecol. Model.*, 173, 295–312.

Tsekeris, T. & Stathopoulos, A. (2006). Gravity models for dynamic transport planning: development and implementation in urban networks. *Journal of Transport Geography*, 14, 152–160.

Tsekeris, T. & Stathopoulos, A. (2009). Real-time dynamic Origin-Destination matrix adjustment with simulated and actual link flows in urban networks. *Transportation Research Record – Journal of the Transportation Research Board*, 1857, 117–127.

Velasco, E. & Roth, M. (2010). Cities as net sources of CO_2: review of atmospheric CO_2 exchange in urban environments measured by eddy covariance technique. *Geography Compass*, 4, 1238–1259.

Walsh, C. J., Oke, T. R., Grimmond, C. S. B. & Salmond, J. A. (2004). Fluxes of atmospheric carbon dioxide over a suburban area of Vancouver. *Preprints for Amer. Meteor. Soc. Fifth Symp. on the Urban Environment*, Vancouver. B. C.

White, R., Engelen, G. and Uljee. I. (1997). The use of constrained cellular automata for high resolution modelling of urban land use dynamics. *Environment and Planning* B, 24, 323–343.

White, R. & Engelen, G. (2000). High-resolution integrated modelling of the spatial dynamics of urban and regional systems. *Computers, Environment and Urban Systems*, 2824, 383–400.

	Planning Alternatives		
	I	II	III
Helsinki	5-storey apartments, 500 residents, minimal impact on green spaces and nature.	5-storey apartments and row of houses accommodating 1,500 residents. Hilltop built; slope non-built.	Residential building around the hilltop all the way down to the waterfront. Office space, maximum 1,000 work-spaces and 1,800 residents.
Athens	Apply cool materials on all buildings at Egaleo municipality and on roads.	Change the land use of Eleonas from brownfield to build area.	Change the land use of Eleonas from brownfield area to green space.
London	Add new street trees.	Add green roofs (varying slopes).	Implementation of both.
Firenze	Complete reforestation of a green and a sport arena in the Cascine Park. Increase of trees (deciduous) by about 75% in total.	Redevelopment of a former industrial area (FIAT) in the north of the Cascine Park, San Donator Park.	Implementation of both.
Gliwice	Sport Centre development.	Centre of new Technologies development.	Implementation of both.

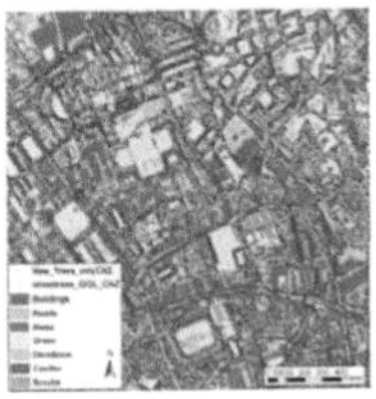

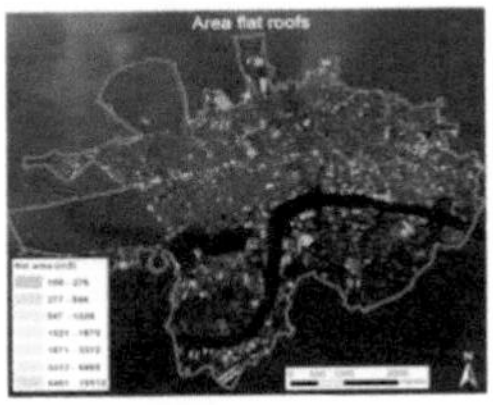

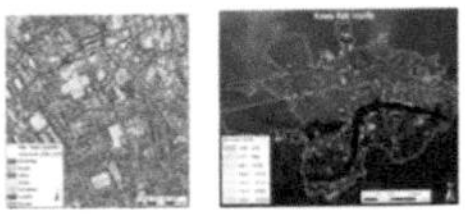

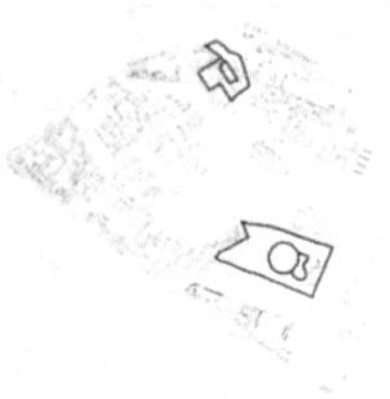

PLATE 1 The urban planning alternatives evaluated within BRIDGE (adapted from Chrysoulakis et al. 2013).

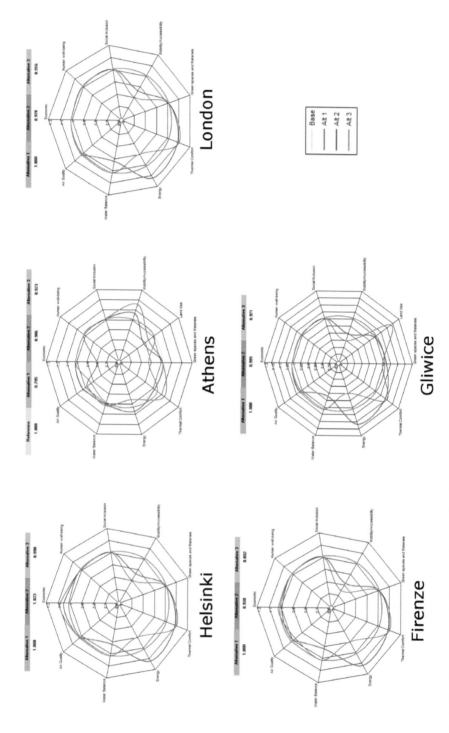

PLATE 2 Spider diagrams and final appraisal scores for all case studies, using the default weight and socioeconomic indicators values (adapted from Chrysoulakis et al. 2013).

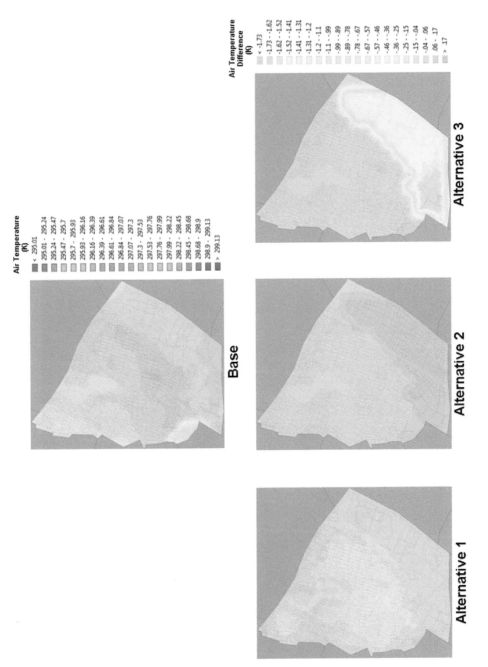

Air Temperature
(K)

- < 295.01
- 295.01 - 295.24
- 295.24 - 295.47
- 295.47 - 295.7
- 295.7 - 295.93
- 295.93 - 296.16
- 296.16 - 296.39
- 296.39 - 296.61
- 296.61 - 296.84
- 296.84 - 297.07
- 297.07 - 297.3
- 297.3 - 297.53
- 297.53 - 297.76
- 297.76 - 297.99
- 297.99 - 298.22
- 298.22 - 298.45
- 298.45 - 298.68
- 298.68 - 298.9
- 298.9 - 299.13
- > 299.13

Base

Alternative 1

Alternative 2

Air Temperature
Difference
(K)

- < -1.73
- -1.73 - -1.62
- -1.62 - -1.52
- -1.52 - -1.41
- -1.41 - -1.31
- -1.31 - -1.2
- -1.2 - -1.1
- -1.1 - -.99
- -.99 - -.89
- -.89 - -.78
- -.78 - -.67
- -.67 - -.57
- -.57 - -.46
- -.46 - -.36
- -.36 - -.25
- -.25 - -.15
- -.15 - -.04
- -.04 - .06
- .06 - .17
- > .17

Alternative 3

PLATE 3 Mean air temperature (K) for the evening period (20:00 – 23:00 LST) for summertime for Athens base case (top) and planning alternatives (bottom), which are maps of differences (adapted from Chrysoulakis et al. 2013).

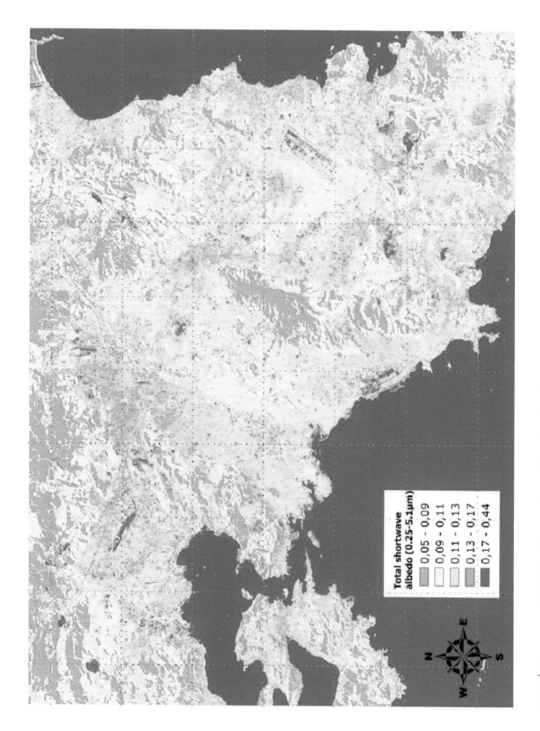

PLATE 4 Shortwave albedo for Athens estimated from Landsat observations.

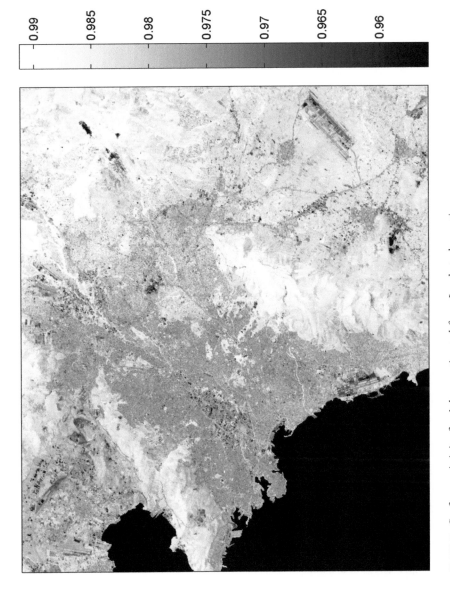

0.99

0.985

0.98

0.975

0.97

0.965

0.96

PLATE 5   Surface emissivity for Athens estimated from Landsat observations.

PLATE 6 The DEM generated for the broader Athens area using ASTER stereo imagery.

Elevation (m)

1411

0

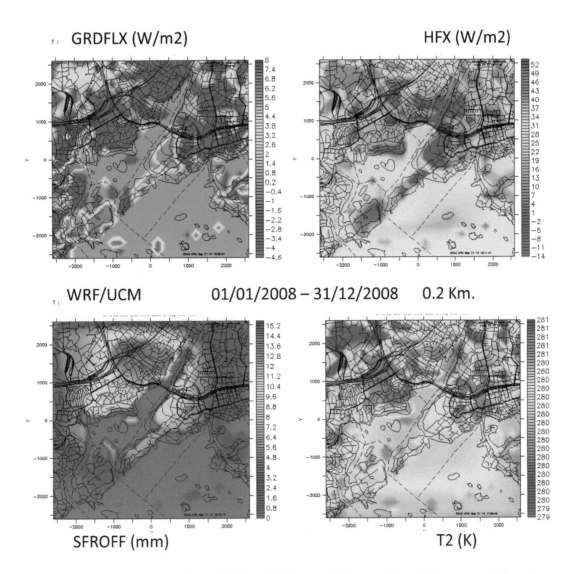

PLATE 7 2008 annual averages for Helsinki at 0.2 km x 0.2 km spatial resolution: ground heat flux in Wm^{-2} (upper-left), sensible heat flux in Wm^{-2} (upper-right), surface runoff in mm (bottom-left) and air temperature in K (bottom-right).

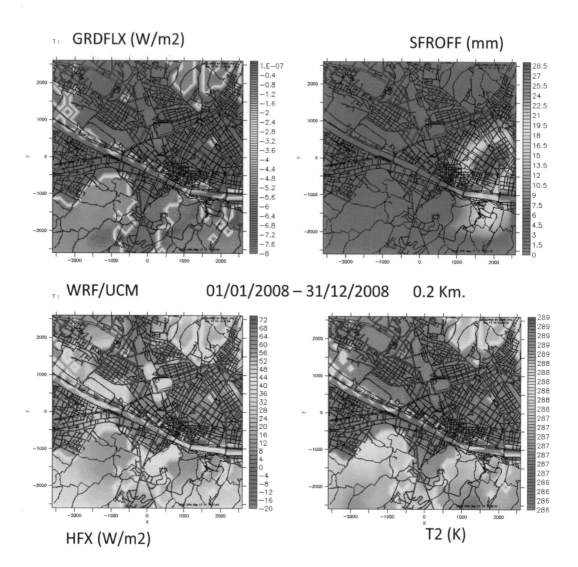

PLATE 8   2008 annual averages for Firenze at 0.2 km × 0.2 km spatial resolution: ground heat flux in Wm⁻² (upper-left), surface runoff in mm (upper-right), sensible heat flux in Wm⁻² (bottom-left) and air temperature (bottom-right).

(a)
(b)
(c)

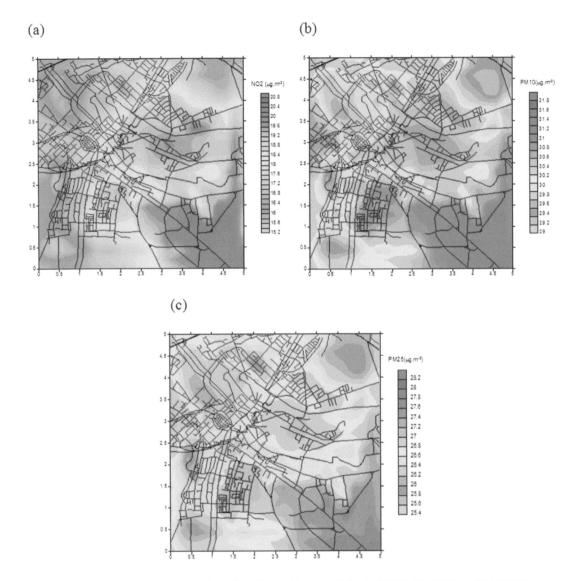

PLATE 9 Annual average concentrations for Gliwice (domain 2) for (a) NO2, (b) PM10 and (c) PM2.5.

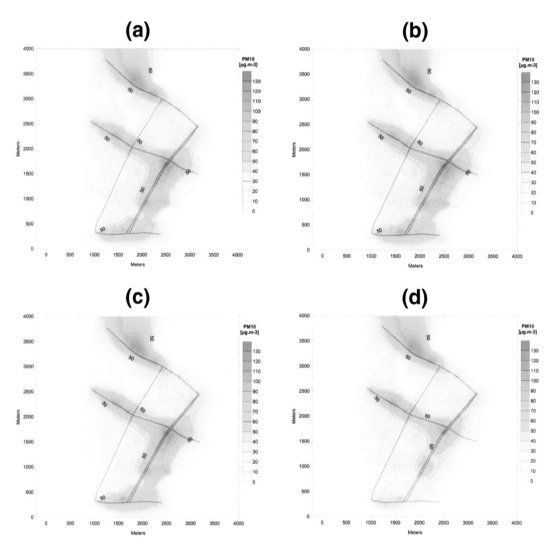

PLATE 10 Comparison of 1.5 m high horizontal fields of 24 hour average PM10 concentration in 22 September 2008 at the Egaleo area in Athens: (a) baseline; (b) PA1; (c) PA2; and (d) PA3 (Borrego et al. 2013).

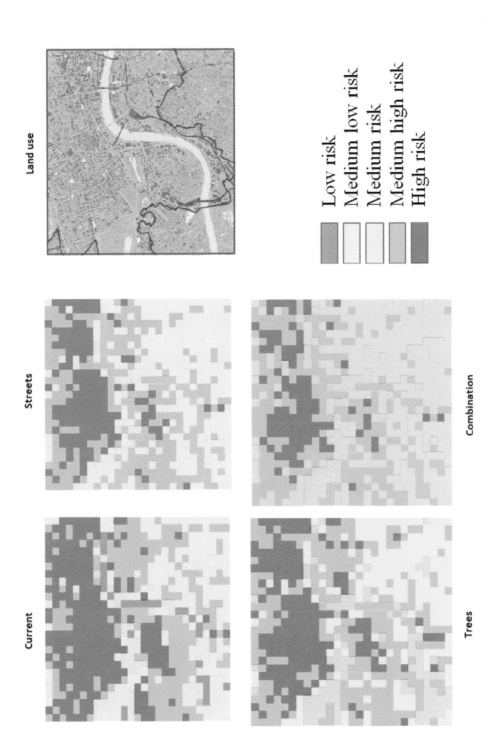

Land use

Streets

Current

Combination

Trees

Low risk

Medium low risk

Medium risk

Medium high risk

High risk

PLATE 11 Spatial differences in risks for higher temperatures in 2010 because of low evaporation rates for the control and the three PA, using the SIMGRO model for the CAZ area in London UK: Current = 70% sealed and 30 % of area green (PA0), Trees = 40% increase of trees (PA1), Streets = 40% more permeable streets (PA2) and combi = combined (PA4). The panel on the right depicts the current main land cover use types: red = built up, green = vegetated area, light blue = open water bodies (e.g. river, lakes, ponds).

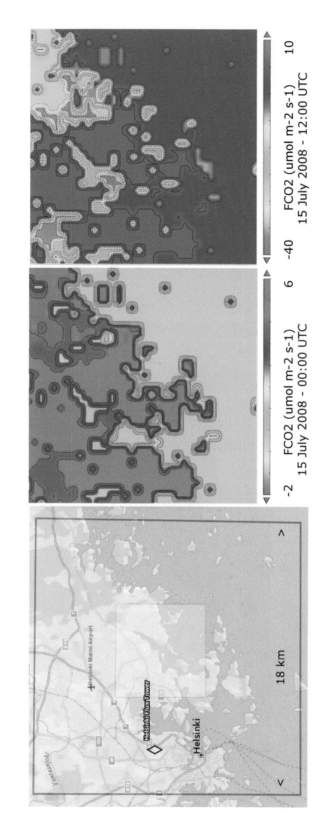

PLATE 12 WRF-ACASA simulation domain (d05, $\Delta x = 600m$) centered on the flux tower area in Meri-Rastila, Helsinki (shaded square) (left panel). The flux tower is in the north of downtown ($\Diamond$). Nocturnal and diurnal maps of CO_2 surface flux (right panels), modified from Falk et al. (in review).

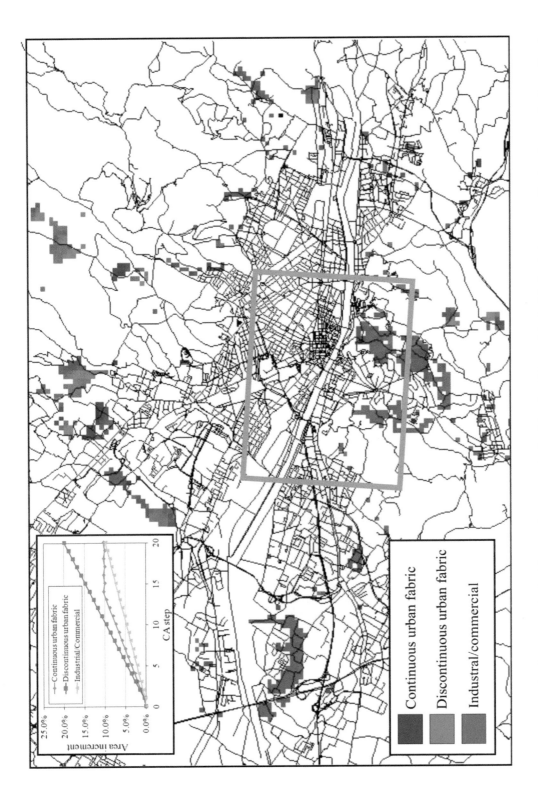

PLATE 13 Future land-use scenario for Firenze city centre obtained by the CA module in 20 simulation steps. Depicted are only the cells where the land use was changed. The embedded graph shows the extensions evolution of the actively modelled land uses (modified from Blecic et al. 2014).

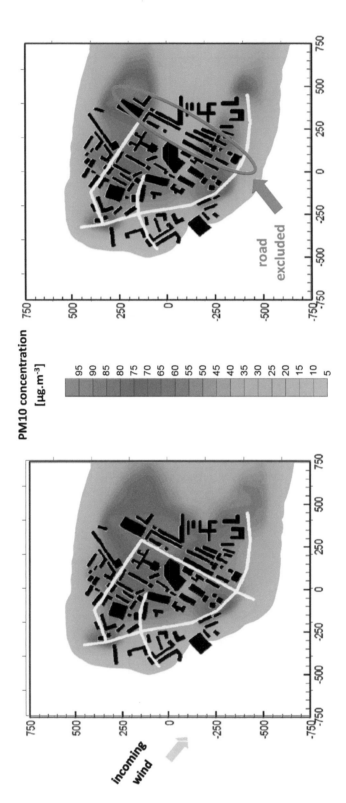

PLATE 14 Example of GIS-based model output for PM10 concentrations, for a road network (left) and for the exclusion of a section of the road (right) in the Helsinki case study.

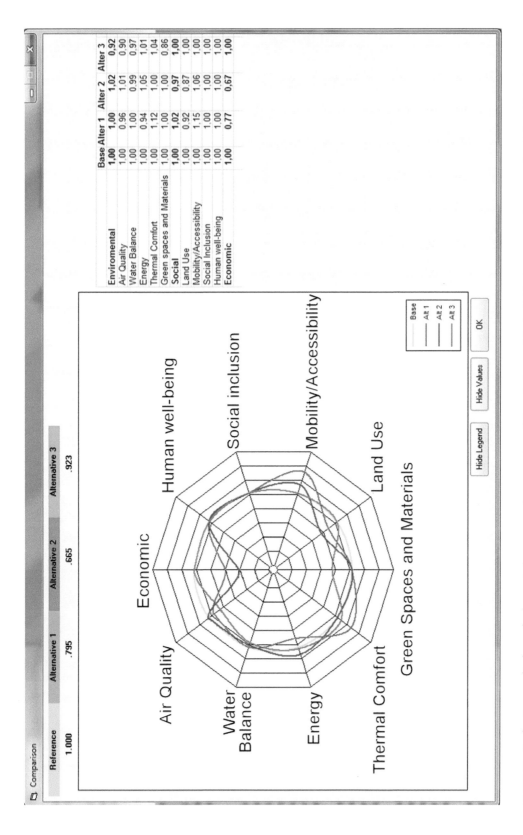

PLATE 15 Evaluation results form: final appraisal scores are shown on the right table and graphically represented in the spider–diagram.

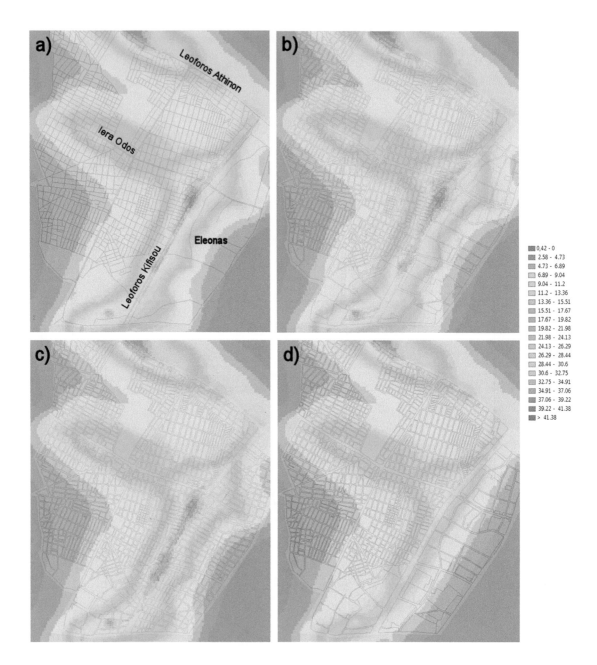

PLATE 16 Indicator maps presenting the distribution of PM10 concentration (μg/m³) in Egaleo (Athens, Greece) for the summer period of 2008 (June – August) for: a) the reference situation, b) the first, c) the second and d) the third planning alternative.

PART III
The socioeconomic components

12

THE USE OF COMMUNITIES OF PRACTICE TO INVOLVE STAKEHOLDERS IN THE DECISION SUPPORT SYSTEM DESIGN

Judith E. M. Klostermann[1], Annemarie Groot[1] and Ainhoa González[2]

[1]ALTERRA - WAGENINGEN UNIVERSITY AND RESEARCH CENTRE AND [2]TRINITY COLLEGE

Introduction: why communities of practice in the BRIDGE project?

The BRIDGE project (Chrysoulakis et al. 2013) aimed to combine scientific models and data from different environmental and socio-economic domains into one integrated Decision Support System (DSS) for sustainable urban planning. The BRIDGE DSS also enabled the incorporation of stakeholder perceptions on the relative importance of the various domains. The purpose of the DSS was to provide useful and timely information to planners and decision-makers on the potential effects of urban planning and development options. Therefore, and in order to ensure the usability of the BRIDGE DSS, end-users had to be involved in the project in a structural way. On-going dialogue between urban planners, DSS programmers, atmospheric modellers, Earth Observation scientists, hydrologists and other technical specialists was thus critical to define the design and outputs of the DSS.

As the concept of a Community of Practice (CoP) was known to be useful in enhancing learning processes (Wenger et al. 2002), the BRIDGE team adopted the CoP approach for structuring the dialogue with stakeholders and the potential DSS end-users. The concept of the CoP was originally developed by Lave & Wenger (1991), who suggested that learning takes place in social relationships rather than through the simple acquisition of knowledge. CoPs are groups of people who share a concern or a passion for something they do and deepen their knowledge and expertise in this area by interacting on an on-going basis (Wenger 1998; Wenger et al. 2002). They are considered a spontaneous, natural phenomenon among people of a similar trade who occasionally meet to learn from each other, and are characterised by three key dimensions: mutual engagement, joint enterprise and shared repertoire.

In addition, there are three critical elements that make up a CoP (Wenger 1998; Wenger et al. 2002):

- Domain: A CoP has an identity defined by a shared domain of interest. Membership implies a commitment to the domain and, therefore, a shared competence that distinguishes members from other people.
- Community: In pursuing their interest in the domain, members engage in joint activities and discussions, help each other and share information. They build relationships that enable them to learn from each other. Having the same job or the same title does not make for a CoP unless members interact and learn together.
- Practice: Members of a CoP are practitioners. They develop a shared repertoire of resources: experiences, stories, tools, ways of addressing recurring problems – in short, a shared practice. This takes time and sustained interaction.

A well-developed CoP that appropriately functions in all three dimensions provides an environment that facilitates learning and knowledge development. A CoP is different from a set of people participating in a stakeholder workshop (that has no continuity) or a social network (that has open boundaries and does not necessarily aim at learning). CoPs are also seen as distinct from other types of groups, such as project teams and working groups, in that they spring from the members' interest and passion, are self-organising, voluntary and have fluid goals around learning.

CoPs fit within an alternative way of thinking about knowledge production that is often referred to as knowledge co-creation (Regeer & Bunders 2009) or transdisciplinary research (Klein et al. 2001). A crucial difference between transdisciplinary research and other forms of research is that the knowledge of local stakeholders/practitioners is considered to be beneficial for the development of sustainable solutions to real world problems. Transdisciplinary research is a response to traditional scientific approaches which have proved to be insufficient to address complex and politically relevant issues, which often cross the borders of sectors and disciplines (Regeer & Bunders 2009; Pohl & Hadorn 2008). Over the years, transdisciplinary approaches have gained importance as a problem-solving approach for fields in which social, technical and economic developments interact with elements of value and culture, including sustainable development, health care and housing (Klein et al. 2001). Arguably, knowledge is acquired and gains meaning within a shared practice. Although the actual use of the CoP as a tool for organising transdisciplinary research has not been scientifically evaluated before, it has been recognised that CoPs can be helpful in facilitating collaboration between researchers and practitioners (Karner et al. 2011).

The CoP approach was introduced in the BRIDGE project in anticipation of an interdisciplinary learning exchange between scientists and planning professionals: the researchers would learn something from the practitioners and practitioners would learn something from researchers. It was acknowledged that both scientists and practitioners were CoP participants with valuable knowledge. The CoP was considered as a learning space where participants could bridge the gap between science and practice (Hearn & White 2009). Informal communications between individuals interested in a given domain become the means for sharing information, improving practice and generating new knowledge and skills (Li et al. 2009). In the case of BRIDGE, communications and information exchange activities were purposefully facilitated through the CoPs. This chapter examines how the CoP approach contributed to the BRIDGE project, and discusses its usefulness in enhancing transdisciplinary learning in research projects, as well as its limitations.

Method: BRIDGE project as case study for CoP application

The BRIDGE project involved five case study cities spread over Europe (See Chapter 3). Two CoP gatherings were organised in each city, and two umbrella CoP meetings were part of the project design as well. The umbrella CoPs provided an opportunity for mutual learning between participants of the five case study cities. Moreover, stakeholders and end-users involved in the project were encouraged to meet before, in between and after the BRIDGE CoP meetings.

The CoPs were organised in parallel to the scientific research work on energy balance, hydrology, CO_2 and pollutants in each city, and they were linked to the different phases of the DSS development. The project design aimed at connecting the CoP meetings to all the other tasks of the project in order to improve the relationship between the CoP organisers, participants and the urban metabolism researchers. Throughout the project, CoP outcomes and participants' perceptions on both the CoP meetings as well as on the project outputs (e.g. urban metabolism indicators, DSS) were gathered and monitored in order to evaluate the performance of the CoP approach and examine its contribution to the project. Interviews were held with BRIDGE researchers and participating practitioners and professionals (e.g. urban planners) to gather their views on the functioning of the CoPs (Garcia-Landarte 2010). Case study practitioners were asked about their general views on the CoP approach (e.g. its capacity to provide inputs into the project and what they could extract from it), as well as about their perceived contribution (e.g. appropriateness of CoP meeting facilitation, relevance of the issues discussed, benefits of participation, usefulness of CoPs in shaping the DSS, contribution of CoPs to enhancing knowledge and changing individual practices, etc.). Similarly, they were consulted on their level of commitment to the CoP. BRIDGE researchers were asked about their level of participation in the umbrella CoP meetings, and their perceptions on the facilitation and usefulness of these meetings, particularly in light of their contribution to the design and development of the DSS. For a full analysis of CoP performance, meeting minutes and official BRIDGE reports were also examined to establish how different participants felt about the approach, how their views changed over time and what effects might be ascribed to the involvement of external end-users through CoP meetings.

Case description: CoPs in the BRIDGE project

It was decided to organise a CoP in each case study city early in the project to initiate and develop communications. To facilitate this process, a CoP coordinator was appointed for each of the cities. Helsinki was the first city that appointed a CoP coordinator, and the rest of the cities followed soon after that. The number and background of participants varied throughout the project; while larger numbers were registered at the first round of CoP meetings, the second round of meetings had a smaller group size (Table 12.1). The umbrella CoP meetings brought together representatives from the five case study cities for a shared debate on the final set of indicators to be incorporated into the DSS as well as for testing the tool.

The first round of CoP meetings was organised and delivered within the first six months of the project. This first set of gatherings had a strong emphasis on building networks and in defining key urban sustainable development issues, which led to preliminary discussions on sustainability objectives and indicators (Table 12.2). During these meetings, the participants suggested specific case study sites and existing project proposals to be analysed by the DSS.

TABLE 12.1 Overview of the first and the second round of CoP gatherings and the two umbrella CoP meetings

CoP1	Date	Participants	CoP2	Date	Participants
Helsinki	15 June 2009	21	Helsinki	20 January 2010	17
London	24 August 2009	24	London	1 April 2010	10
Athens	8 October 2009	50	Athens	18 February 2010	29
Florence	16 October 2009	17	Florence	3 December 2009	14
Gliwice	20 October 2009	30	Gliwice	28 January 2010	26

Umbrella CoP1		Date		Participants	
Athens		5 May 2010		22	

Umbrella CoP2		Date		Participants	
Aveiro		17 March 2011		35	

Final BRIDGE Demonstration Meeting		Date		Participants	
Brussels		26 October 2011		52	

TABLE 12.2 Objectives and indicators and their development during the BRIDGE project (adapted from González et al., 2010a, 2011; Breil et al., 2010)

Objectives defined in first CoP meetings	*Indicators proposed in second round of CoP meetings*	*Final list of indicators agreed at the umbrella CoP meeting*
Improve air quality and reduce greenhouse gas emissions	• Concentration of pollutants: NO_x, SO_x, PM_{10}, $PM_{2.5}$, O_3, CO ($\mu g/m^3$) • Spatial distribution of pollutants: PM_{10}, $PM_{2.5}$, NO_x, SO_x, CO_2 ($\mu g/m^3$) • Relationship between pollutant concentrations and wind direction • Number of days above established air quality thresholds • Effects of meteorological conditions (e.g. temperature) on pollutant concentrations • Greenhouse gases and CO_2 emissions (tons) • CO_2 emissions (tons or % of reference value) • Source of emissions (% per building/sector type) • Contribution from single boilers to total emissions (%) • Emissions from transport and split per type – private and public (%)	• Concentration of pollutants: PM_{10} and $PM_{2.5}$, O_3, NO_x ($\mu g/m^3$) • Greenhouse gases and CO_2 emissions (tons or % of reference value) • Number of days above established air quality threshold (days)
Improve energy efficiency	• Energy demand (electricity consumption kWh/m^2 or % change) • Heating demand (kWh/m^2 or % change) • Energy consumption for lighting the avenue (kWh/m^2) • Urban temperature outdoors – compared to rural temperatures (°C) • Percentage of energy from renewable sources such as solar panels (%) • Potential renewable energy from the volume of biomass produced • Percentage of energy created in a decentralized way (%) • Additional heat generated in a decentralized way (W/m^2) • Percentage and structure of thermo-insulation (%) • Energy balance in buildings (i.e. energy heating)	• Energy demand (kWh/m^2) • Potential for renewable energy (type of renewable sources) • Additional heat generated (W/m^2) • Percentage of energy created from renewable resources (%)
Anticipate climate change (flooding)	• Flooding zones (location and extent in m^2) • Number and extension of flood risk areas (location and extent in m^2)	• Flooding zones and risk areas (location and extent in m^2)
Optimize water use and management	• Urban water supply and use (m^3/capita) • Water balance: surface run-off, evapotranspiration, and filtration (mm^3/m^2) • Volume of water used for irrigation (m^3) • Percentage of wastewater treated and houses connected to wastewater treatment plant (%)	• Surface run-off, evapotranspiration and filtration (mm^3/m^2) • Water consumption per capita (m^3)

Objectives defined in first CoP meetings	Indicators proposed in second round of CoP meetings	Final list of indicators agreed at the umbrella CoP meeting
Increase green space areas	• No indicators proposed★	• Density of green areas (m^2/m^2 or % of total) • Canopy/green surface or area newly created (m^2) • Accessibility to green areas (no. inhabitants within 500m of green area)
Improve thermal comfort	• Average outdoor temperature (air) and humidity (°C) • Ambient temperature at 1 m above street level (°C) • Average surface temperature in roads, buildings, etc. (°C) • Wind speed (m/s) • Number of days above 33°C /per area (heat waves)	• Ambient and surface air temperature (°C) • Number of days above established threshold (days)
Optimize materials used	• No indicators proposed★	• Volume of material re-used/ recycled (m^3 of total)

★ In such cases, end users did not propose any indicators but the BRIDGE team subsequently proposed relevant indicators based on international literature and best practice.

During the second round of CoP meetings planning objectives, criteria and indicators, including socio-economic criteria, were defined for each of the case studies (González et al. 2010a). The participating groups were smaller, with more focused expertise, and the interaction between scientists and end-users more intense.

The first umbrella CoP meeting was organised after the second round of CoP meetings at city level. Each of the five case studies was represented by one or two participants at this meeting (Table 12.1). Participants shared their views on urban planning practices, and discussion on the proposed indicators led to an agreement on the final set of core (i.e. common to all case studies) and discretionary (i.e. specific to the case studies) indicators to be incorporated into the DSS. The beta version of the BRIDGE DSS was presented to the participants to provide an idea of its scope and functionality. CoP participants (i.e. planning practitioners and DSS end-users) were asked about their expectations of the DSS. Overall, participants considered that the DSS should be a learning tool; they wanted to be able to understand the underlying models and scoring system framework, and they wanted a fast and user-friendly tool to explore planning proposals.

The second umbrella CoP focused on scenario development and analysis (see Chapter 17). A number of future strategic scenarios was presented to the participants (Marques et al. 2010), who then attributed importance weights to the agreed objectives and indicators based on such scenarios (González et al. 2010b).

A final project seminar was organised to demonstrate the final draft of the DSS prototype. BRIDGE researchers, case study practitioners, European Commission representatives and interested professionals participated in this final event where DSS testing enabled further gathering of stakeholder perceptions on the usability and user-friendliness of the tool. This, in turn, enabled fine-tuning the DSS for its final delivery.

Results of the interviews

The results from the two sets of interviews among the BRIDGE researchers and end-users have been grouped into the three aspects that make out a CoP: the domain, the community and the practice.

Domain

When asked for a definition of sustainable urban planning, the majority of respondents used broad definitions, including the classical aspects of sustainability: ecology, economy and social welfare. In all cases, there was a broad agreement on this concept which suggested the existence of a shared domain for the CoP participants. However, identified sustainability issues were different, depending on the urban context and the respondents' practice and research disciplines. Examples included social and economic aspects and climate change in London, urban sprawl and zero carbon in Helsinki, transport systems and brownfield land in Gliwice, and air pollution in Florence. Interview responses about the urgency of sustainable urban planning present further evidence that sustainable urban planning is a shared domain that connects all the participants.

According to CoP theory, the discussions and learning should follow the agenda set by the participants themselves. However, in the case of BRIDGE, the BRIDGE team largely determined the agenda for the CoP meetings. Although the broad subject of sustainable urban planning was a shared domain, the project scope determined which topics within this domain were to be addressed (e.g. air quality and CO_2 were more important than social cohesion or urban sprawl). To promote self-organising CoPs, the meetings were arranged in an open manner. In all cases, draft agendas were circulated to case study coordinators and invited speakers for their review and feedback. Nevertheless, end-users felt that their input in the CoPs could have been more effective if they had had more influence on the meeting agenda.

When asked what content was discussed during the CoP meetings, most respondents highlighted the indicators. This was a concrete outcome that the BRIDGE project needed from the CoP meetings: an overview of the important environmental and planning themes in each city, and the objectives and the indicators to measure progress towards those objectives. Another important topic was how the BRIDGE DSS would work (e.g. ensuring that it was interactive and that it visualised the impact of planning decisions). Respondents also reported to have learned a number of other things:

- How planning systems in the different countries work; for example, that procedures can be different and complex, but that the planning decisions are market-driven everywhere.
- How different cities in Europe are working to solve their economic, social and environmental problems; for example, through realising new urban green areas.
- What methods and models were used by the BRIDGE team; for example, the foresight method and the urban metabolism approach including energy and material flows.

Community

All respondents indicated that they met new people during the CoP meetings. This was considered very valuable particularly in the context of the cities. BRIDGE brought local experts together from different fields that would not have sought each other out in normal circumstances. In many cases, these were the contacts that continued once the BRIDGE CoPs ended. Meeting planners from other cities was another valuable experience for the participants; they learnt from each other during those interactions but, in general, the contacts were not continued after the project ended.

The BRIDGE researchers working in the case study cities came into contact with practitioners in their city or strengthened existing contacts. This led to the most concrete continuation of contact: they exchanged data, and talked about new projects. Several BRIDGE researchers commented that they were sceptical about CoPs in the beginning, but that it was rewarding in the end; their commitment to the CoPs increased over time. Networking with urban planners and seeing how they operate were the perceived added value of the CoP meetings. The outcome that new contacts were an important benefit for CoP participants indicates that there were good grounds for starting CoPs at the city scale.

The BRIDGE team participated in the CoP because it was part of the scope of the project. In contrast, the urban planning representatives participated because they were interested in new methods for urban planning. Although there was relevant exchange and learning, the fact that the DSS was not ready until the end of the project was a significant limitation according to these participants. Difficulties in data gathering and DSS development impeded a fully effective exchange.

The time invested by interviewees for one CoP meeting varied from half a day to a couple of days. BRIDGE researchers with an organising role spent most of their time in meeting preparations. The time invested by the end-users was often quite significant, in the order of a couple of days, as they held meetings in preparation of the CoP meeting. Monitoring during the CoP meetings indicated that both groups (i.e. BRIDGE researchers and case study representatives) participated actively, with equal opportunities to intervene and learn from each other.

According to the interviewees, there is a gap between scientists and urban planners, and the CoP played a role in bridging this gap. Nevertheless, according to some urban planners, the gap remains. BRIDGE scientists followed the pre-designed project plan, without taking the needs of planners fully into account. A suggested solution was involvement of planners/stakeholders at the project proposal stage. In contrast, the BRIDGE researchers considered that end-users should have been involved at a later stage, when the BRIDGE DSS was operational. Overall, although it is difficult to establish whether urban planners should have been involved earlier or later in the process, a clear effort was made to incorporate the needs of end-users into the project.

None of the case studies has continued the CoP meetings after the BRIDGE project completion. Some interviewees showed a willingness to participate if there was a follow-up, but nobody felt responsible for organising it. Nevertheless, individual contacts do continue. In some cases, the CoP approach has strengthened existing networks and, in others, it has led to new networks. One of the difficulties for their continuity was the heterogeneity of the BRIDGE CoPs; despite the fact that all participants had an interest in sustainable urban planning, their roles, daily routines and cultures were very different. The organisation of such a diverse meeting was a significant effort. Having said that, the heterogeneity of the group was one of the main attractions of the BRIDGE CoPs. Therefore, it can be concluded that the effort of organising the CoP meetings was considered the main bottleneck for their continuity.

Practice

All the interviewees thought that the CoP received enough attention within the project. A fair amount of time was spent discussing relevant aspects with the end-users. The main influence the end-users had on the project was in defining the DSS indicators. Although many of the BRIDGE researchers made an effort to consult and take into account the feedback of the end-users, another group of BRIDGE researchers was not involved in the CoPs. The structure of the project was too rigid for them to really have an open discussion with the end-users. Nevertheless, the practice of the BRIDGE researchers was changed by the influence of the urban planners, e.g. through an increased awareness of the needs of end-users of scientific projects, and of the connectedness of land-use planning, energy flows, air pollution and water consumption.

The BRIDGE researchers did not fully succeed in changing the practice of the urban planners. The interviewees noted that not having seen a fully working DSS was the key limitation. The DSS was considered to be a potentially very valuable tool as the process of urban planning is in dire need of a more systematic approach to sustainability. All the urban planners emphasised that an operational DSS would definitely have changed their practices. Interviewees also noted, among other things, that:

- The BRIDGE researchers did not model the process of the DSS after planning practices so it would be difficult to apply.
- Monitoring and data collection is different in the different cities which makes it difficult to design one tool for Europe.
- Specific issues that are relevant for urban planning such as health, urban sprawl and socio-economic factors were not incorporated.

It can be concluded from the interview results that participants considered the DSS to have significant potential in supporting urban planning, but this was difficult to realise within the project time frame.

Conclusion: is there a future for CoP as a part of transdisciplinary projects?

A CoP functions where there is a shared domain, a community and a shared practice (Wenger 1998). From the analysis above, it can be concluded that there certainly was a shared domain, namely that of sustainable urban planning. For researchers as well as urban planners this was an urgent issue and to learn more about this was an important motivation to participate in the CoPs. The agenda, however, was dominated too much by the BRIDGE project goals, which reduced the learning opportunities for the urban planners.

There were difficulties in developing the BRIDGE community. Meeting people from different organisations and backgrounds was seen by the respondents as very valuable and an important benefit of their participation in the CoPs. However, the CoP was an external initiative in each city and not a 'natural' process of people wanting to meet each other. Moreover, the CoP was developed in the context of a project so the effort to organise meetings ended with the project. The differences in organisational goals (for example, models and methods versus practical solutions) and the differences in culture and in geographical location of the participants also created barriers for the continuation of the CoPs. The organisation of the CoPs took a lot of effort and the rewards were apparently not significant enough for anyone to follow it up after the BRIDGE project ended. What remained were individual contacts between some of the most active CoP participants.

Concerning the changes in the daily practice, the BRIDGE researchers benefitted more from the CoP meetings than the urban planners. The CoPs provided input for the DSS in the form of urban sustainability issues, objectives, indicators and associated importance weights, as well as guidance on the design of the DSS and its outputs. The CoP outcomes were actually used in their work, and their interest in the CoPs increased during the project. The urban planners, on the other hand, became gradually disappointed because there were only a few practical outcomes from the project that they could immediately use. They felt that the project should have been more flexibly formulated so that their input could influence the DSS design in a more fundamental way. Nevertheless, the end-users shaped the DSS by incorporating their preferences in the indicator set. End-users also required the DSS to provide mapped and tabular results for each of the indicators considered rather than a final performance score for a given alternative as the BRIDGE team had planned. With more detailed outputs end-users could examine the environmental effects of planning proposals more effectively. Limitations were encountered when trying to address the socio-economic concerns of urban planners (e.g. employment and health) due to the lack of socio-economic models within the BRIDGE project. The DSS was not fully operational until the end and there were questions as to where the tool would fit in the planning process. Had the DSS been fully operational, the practice of the urban planners would definitely have changed as there was a clear need for such a tool.

It can be concluded that a good effort was made to create a platform of exchange between scientists and practitioners and that the meetings provided a structured platform for learning and knowledge exchange. Although the effectiveness of CoP interactions could have been improved, they led to a significantly larger end-user input, compared to many other similar projects. The differences in cultures and goals prevented the emergence of a real, voluntarily functioning CoP in the five case study cities. Nevertheless, the benefits for the project were obvious and were noticed by the European Commission's project reviewers, who stated that they had seldom seen a scientific project where end-users have so much influence at such an early stage in a multidisciplinary project.

It could be argued that the BRIDGE network qualified as a 'network of practice' (Brown and Duguid 2000) which is made up of people that engage in the same or very similar practice, sharing a great deal of common practice and implicit understanding. The sharing of ideas is achieved through less direct communications such as professional newsletters, mailing list servers, journals and conferences.

We conclude that it is difficult to develop a CoP in a multidisciplinary project with many strictly defined deliverables and a tight timetable. Early and prolonged interaction with the end-users in a project-based context is in itself a good idea; however, this does not have to qualify as a CoP but rather could be labelled an end-user panel or focus group. If a CoP grows out of such an effort in a 'natural' way, so much the better. A project totally dedicated to setting up a CoP on sustainable urban planning in municipalities is also a good idea, but it should not be overburdened with the demands of one scientific project. In such a dedicated CoP, the frequency of meetings can be higher and the agenda setting can be more open and 'democratic'.

References

Breil, M., Mysiak, J. & González, A. (2010). Indicators definition report. *BRIDGE Deliverable 5.3*. Online. Available HTTP: <http://www.bridge-fp7.eu>.

Brown, J.S. & Duguid, P. (2000). *The social life of information*. Boston, MA: Harvard Business School Press.

Chrysoulakis, N., Lopes, M., San José, R., Grimmond, C.S.B., Jones, M.B., Magliulo, V., Klostermann, J.E.M., Synnefa, A., Mitraka, Z., Castro, E., González, A., Vogt, R., Vesala, T., Spano, D., Pigeon, G., Freer-Smith, P., Staszewski, T., Hodges, N., Mills, G., & Cartalis, C. (2013). Sustainable urban metabolism as a link between bio-physical sciences and urban planning: the BRIDGE project. *Landscape and Urban Planning*, 112, 100–117.

Garcia-Landarte, D. (2010). Monitoring Communities of Practice in the context of BRIDGE. MSc thesis, ALTERRA, Wageningen UR, The Netherlands.

González, A., Donnelly, A. & Jones, M. (2010a). Socio-economic and environmental workshops report. *BRIDGE Deliverable 5.1*. Online. Available HTTP: <http://www.bridge-fp7.eu>.

González, A., Donnelly, A. & Jones, M. (2010b). Report on the impact assessment model for urban metabolism. *BRIDGE Deliverable 5.2*. Online. Available HTTP: <http://www.bridge-fp7.eu>.

González, A., Donnelly, A., Jones, M., Klostermann, J., Groot, A. & Breil, M. (2011). Community of Practice approach to developing urban sustainability indicators. *Journal of Environmental Assessment Policy and Management*, 13(4), 591–617.

Hearn, S. & White, N. (2009). Communities of Practice: linking knowledge, policy and practice. Online. Available HTTP: <http://www.odi.org.uk/sites/odi.org.uk/files/odi-assets/publications-opinion-files/1732.pdf>.

Karner, S., Rohracher, S., Bock, B., Hoekstra, F. & Moschitz, H. (2011). Knowledge brokering in Communities of Practice: synthesis report on literature review. Foodlinks project – Knowledge brokerage to promote sustainable food consumption and production: linking scientists, policymakers and civil society organizations. Online. Available HTTP: <http://www.foodlinkscommunity.net/>.

Klein, J.T., Grossenbacher-Mansuy, W., Haberli, R., Bill, A., Scholz, R.W. & Melti, M. (eds). (2001). *Transdisciplinarity: joint problem solving among science, technology and society—an effective way for managing complexity.* Birkhauser Verlag, Basel.

Lave, J. & Wenger, E. (1991). *Situated learning – legitimate peripheral participation.* Cambridge University Press, Cambridge.

Li, L.C., Grimshaw, J.M., Nielsen, C., Judd, M., Coyte, P.C. & Graham, I.D. (2009). Use of Communities of Practice in business and health care sectors: a systematic review. *Implementation Science,* 4 (27), doi: 10.1186/1748-5908-4-27.

Marques, M., Wolf, J., Borges, M., Lopes, M., Martins, J.M. & Mar, H. (2010). Strategic scenario analysis. *BRIDGE Deliverable 7.1.* Online. Available HTTP: <http://www.bridge-fp7.eu>.

Pohl, C. & Hadorn, G.H. (2008). Methodological challenges of transdisciplinary research. *Natures Sciences Sociétés,* 16, 111–121.

Regeer, B. & Bunders, J. (2009). *Knowledge co-creation: interaction between science and society – a transdisciplinary approach to complex societal issues.* Free University Amsterdam, Athena Institute, The Netherlands.

Wenger, E. (1998). *Communities of Practice: learning, meaning, and identity.* Cambridge University Press, Cambridge.

Wenger, E., McDermott, R. & Snyder, W.M. (2002). *Cultivating Communities of Practice: a guide to managing knowledge.* Harvard Business School, Boston, MA.

13

COLLECTION OF SOCIO-ECONOMIC DATA AND INDICATORS FOR URBAN INTEGRATED MODELLING

Margaretha Breil[1] and Ainhoa González[2]

[1]CENTRO EURO-MEDITERRANEO PER I CAMBIAMENTI CLIMATICI S.C.A.R.L. AND [2]TRINITY COLLEGE DUBLIN

Introduction: why use socio-economic indicators?

Interventions in urban areas, further to causing changes in the urban metabolism, directly influence living environments of the population, including the environmental and economic aspects of living conditions. This interconnectedness between physical environment, economic development and social well-being is deeply anchored in the concept of sustainability, as defined in the seminal statement by G.H. Brundtland, stating that sustainability ' . . . is not only a new name for environmentally sound management, it is a social and economic concept as well' (Brundtland 1987).

The concept of urban metabolism, itself, implicitly includes the consideration of inter-relationships between economic, social and environmental living conditions and hence also addresses urban sustainability. The definition provided by Kennedy et al. (2007) which describes urban metabolism as ' . . . the sum total of the technical and socio-economic processes that occur in cities, resulting in growth, production of energy, and elimination of waste . . . ' explicitly goes beyond the accounting for physical flows connected to urban areas, and also considers the social and economic sphere. Thus, if the concept of urban metabolism is used as criteria for the assessment of planning alternatives, the concept needs to take into consideration, alongside physical criteria like use of natural resources and production of waste, ' . . . the improvement of its liveability, so that it can better fit within the capacities of the local, regional and global ecosystems' (Newman 1999). Therefore the assessment framework underlying the development of the Decision Support System (DSS) in the framework of BRIDGE project (Chrysoulakis et al. 2013) (see Chapter 16 for details) required the consideration of potential changes in urban economic and social conditions affecting human well-being, and needed to include appropriate indicators which capture and assess these aspects, too.

Measurement of social and economic well-being

Methods for the measurement of human well-being have been widely discussed since the publication of several reports on alternative options for measurement of issues of personal and societal well-being and social justice, especially since the publication of the report on the 'Measurement of Economic Performance and Social Progress' presented to the French government by Stiglitz et al. (2009a,b). Alongside this report, a series of approaches have been developed that go beyond the traditional metrics of sustainable development, where mainly Gross Domestic Product (GDP) and income are used for the assessment of social progress. The discussion following the publication of the Stiglitz report has succeeded in broadening consensus for qualitative, non-economic forms of measurement to better capture relevant elements of human well-being and thus accounting for aspects that go beyond criteria connected mainly to the dimension of economic growth (Markandya & Tamborra 2006; Carraro et al. 2013). Further attempts to operationalize the principle of sustainability for urban policies were made in local Agenda 21 efforts, in the sustainable cities movement and in the urban sustainability indicator set used by the European Environmental Agency. The indicator sets developed include indicators based on quantitative aspects (measurements of material flows, as for instance, waste quantities and CO_2 emissions) and qualitative indicators based on population satisfaction (Table 13.1). In all cases, extending the criteria of sustainability to aspects of personal well-being aims at introducing the consideration of these concerns into the public debate on physical transformations.

Nevertheless, the consideration of social and economic dimensions for the definition of sustainability indicator sets encounters obstacles in terms of availability of appropriate and meaningful data. Whereas quantitative statistical data describing social and economic dimensions of sustainable urban metabolism, for instance with regards to demographics or income, is commonly available, the measurement of social well-being requires either the use of information from qualitative surveys or from expert knowledge; in any case this data will necessarily have a more qualitative than quantitative character.

The role of indicators for sustainability appraisal

The concept of sustainability applied to urban contexts requires the consideration of quite complex interrelations and of a great variety of aspects. In order to limit the amount of information to be assessed

TABLE 13.1 European common indicators for urban sustainability

Indicators	Units
Citizen satisfaction at the local level	Personal security (% of satisfied population)
	Natural environment (% of satisfied population)
	Urban environment (% of satisfied population)
	Services (culture and arts) (% of satisfied population)
Local contribution to climate change	CO_2 emissions per capita per sector
Mobility and transport	% of private car use and % of other transport modes
Accessibility to green areas and services	% of population that lives 300m away from a green area >5,000m²
Air quality	Net overexposure to PM10
Transport of kids	Home-school travel: % of kids brought to school by private car
Sustainable management of local authorities and local enterprise	% of environmental certificates in relation to total number of businesses (per business type).
Noise impacts	% of population exposed to noise night levels L>55dB(A)
Sustainable use of land	% of protected/urbanized areas in relation to the total of the administrative area
Sustainable products	% of sustainable products acquired
Waste	% of waste produced and type

Source: EEA 2005.

and to ease the communication process between all parties involved, data is normally introduced in the form of indicators, which can meaningfully represent the underlying more complex issues and related trends (EEA 2005). Indicators which are able to 'translate' complex into 'manageable' (Moussiopoulos et al. 2010) but still hold meaningful information are generally appreciated as useful tools for communicating among the different actors involved in decision making, including the public. Their use is of central importance given their role in informing decision making, as well as in monitoring of policy implementation (Valentin & Spangenberg 2000). They have been widely used for assessments of development of urban areas towards concepts like 'sustainability' or 'resilience', and have been particularly successful in terms of making achievements comparable across different urban contexts (Shen et al. 2011). If implemented in a context of multi-criteria decision making like the one of the BRIDGE DSS, they provide a framework for addressing trade-offs and individual preferences with respect to heterogeneous criteria and preferences.

Sustainability represents a goal rather than an achievement to be defined and measured in absolute terms. Therefore, the use of this concept as a criterion for evaluating or scoring alternative urban transformation options implies that an assessment, rather than providing judgements on sustainability of the urban environment in absolute terms or as trends, needs to capture the differences in the contribution of each alternative to the overall goal of increased sustainability of the urban metabolism.

Sustainability appraisal in the context of BRIDGE

Furthermore, unlike the common use of sustainability indicators as a means for monitoring on-going policies, the BRIDGE DSS aims at assisting integrated assessments for sustainable urban planning decision making; thus BRIDGE indicators cannot rely on existing data, but require data from modelling and projections in order to be able to anticipate future impacts. With regards to the physical aspects of urban metabolism, the BRIDGE approach aims at employing detailed quantitative modelling in order to provide information on impacts expected from different planning alternatives (González et al. 2013). On the contrary, impacts connected to human well-being, quality of life or economic perspectives of inhabitants cannot be based on model outputs, requiring alternative data sources. The retrieval of data on social and economic impacts of the interventions to be assessed encountered a series of challenges:

- The relatively small scale of interventions considered for the application of the BRIDGE DSS created problems in terms of scale as, frequently, data was difficult or impossible to retrieve at a sufficient spatial detail.
- Relating impacts expected from small scale interventions to the entire city area furthermore presented problems in terms scale mismatch, as up-scaling of impacts to higher planning levels or wider geographical extents are difficult to perform accurately and may generate outputs that are either difficult to appreciate, or not useful for the assessment of planning alternatives.
- Indicators needed to consider interactions between different spatial scales; however, potential impacts and trade-offs between trends within the planning area and in other parts of the same city are difficult to assess focusing exclusively on reduced areas, particularly if the more complex mechanisms of housing and real estate markets are concerned.
- The quantification of large scale future economic impacts, albeit theoretically feasible, was beyond the scope of the project.

For these reasons, the consideration of qualitative data and expert judgements alongside quantitative data from model outputs became an imperative for the completion of the indicator sets used for the

BRIDGE DSS. The mixed approach of integrating quantitative with qualitative criteria must be seen as an option offering considerable advantages against purely qualitative or quantitative approaches. These advantages go far beyond the question of information availability: on the one hand, quantitative approaches to the definition of sustainability indicators can be seen as privileging 'technique over reflexively engaging people' (Scerri & James 2010), giving room to the creation of knowledge about sustainability, rather than the mere transmission of information; on the other hand, qualitative approaches allow for the discussion of values and norms related to the definition of sustainability criteria in local contexts and about the relative importance of each contribution. The advantages of using qualitative indicators lay in their potential to reflect human interaction and stakeholders' subjective assessments and thus 'make a greater contribution to understanding and practicing sustainability' (Scerri & James 2010).

In this sense, some of the socio-economic indicators included in the BRIDGE DSS have been designed as 'placeholders' to be filled, for every new application of the DSS, with locally significant indicators to be determined in discussions with stakeholders in the specific local contexts. In doing so, the flexibility and potential adaptability of the DSS tool has been increased.

Choice of data and indicators

Among the indicators chosen for the BRIDGE DSS which address specific aspects of urban quality and of quality of life in a bottom up procedure, those representing areas of particular concern have been defined during the CoP meetings in each case study city (see Chapter 12). In a subsequent phase, these indicators have been harmonized by choosing areas of concern articulated in more than one case study site and variables more widely adopted in similar exercises; see for instance the indicator sets proposed in Provincia di Firenze (2005), Comune di Firenze (2008) and European Commission (1999). This was the case for instance of social and economic criteria addressing impacts on the local economy and mobility. Nevertheless, even data concerning the European common indicators set (Table 13.1), such as use and production of sustainable products, were not readily available for all urban areas. The alternative strategy adopted was to draw back on expert judgements. Rather than producing metric scales, the qualitative judgements focus on the relative differences between impacts expected from planning alternatives, creating ordinal scales: for instance, assessing that alternative A will have a greater positive economic impact (i.e. job creation opportunities) than alternative B. The blending of these rather descriptive indicators and those based on quantitative criteria (Holden 2006) points to the crucial role the choice of indicators and their weighting have in decision making processes (Olewiler 2006). The BRIDGE DSS tool, further to integrating data inputs based on expert judgements, allows for the re-interpretation of criteria and indicators proposed in participative processes.

Indicators for assessing and monitoring urban development proposals

In the context of sustainability assessments, indicators are a commonly used tool for communicating among different actors involved in decision making, including the public, as they 'translate' complex information into 'manageable' and meaningful information (Moussiopoulos et al. 2010). The effective transfer of information and generation of knowledge through the use of indicators are commonly achieved by a simplification of the observed reality. Therefore, the collection of data, in addition to availability constraints, is focused on the most relevant aspects and on the attempt to provide valuable information which adds significance to what is obtained from directly measured parameters or properties (Smeets & Weterings 1999). For example, an indicator measuring a single pollutant can address some aspects of the more complex air pollution problem, as well as capture other associated issues such as land use change. Similarly, an indicator can combine the measurement of two or three single values (e.g. urban density, rate of green areas urbanized and use of brownfields for new developments) illustrating the overall land use change trends. In order to be meaningful and useful for decision makers, scientific data is therefore commonly translated into indicators, in some cases grouped according to criteria based on policy goals (i.e. aggregated into a simple or composite value, which enables the identification of trends correlated to the policy action under assessment).

Several indicator sets have been developed and are currently available at the European Union (EU) and national level (EC 2003; EEA 2005; Bardos et al. 2009). We can refer, in particular, to the core set of environmental indicators developed at European level (Smeets & Weterings 1999; EEA 2005), the European Common Indicator Set (EC 2003) and the local-level data gathered through Agenda 21 initiatives; see, for instance for Firenze, one of the case study cities (CDF 2008). These indicator sets have been designed explicitly for sustainability monitoring procedures, and have generated significant data collections and valuable shared practices, experiences and knowledge in accounting for urban sustainability.

Within the scope of the BRIDGE project, relying on existing consolidated datasets obviously presented a series of advantages associated with consistent and periodically available measurements, as well as with comparability across different contexts – valuable aspects which are normally not provided by ad hoc locally defined indicator sets. However, ad hoc indicators tailored to the specific characteristics of a planning proposal, or concern have the potential to more precisely measure and to more accurately address the problem at hand. This dichotomy between generic and case-specific indicator sets represents a common problem when addressing sustainability at the local level. This is apparent from the more specific national indicator sets that followed European initiatives, which were adapted to specific priority policies, objectives and targets (e.g. EPA 2008; DEFRA 2009; FNCSD 2009).

In the context of BRIDGE, this dichotomy becomes even more critical, given the objective of defining a practical tool to support decision making which can be applied to different urban contexts and decision making problems. The need to define an indicator set that captures local concerns, while being transferable across Europe, represents thus a significant challenge. This is recognized by Alberti (1996), who notes that there is no single set of sustainability indicators applicable to all urban environments as progress towards urban sustainability necessarily has different action needs in different urban contexts. Therefore, the composition of indicator sets, and the relative importance or weight attributed to each indicator during the sustainability assessment, are context-specific and need to address the specific local planning issues. They should focus on relevant environmental and socio-economic priorities and objectives, providing an adequate level of detail for the assessment and ensuring measurability or the possibility of scoring. In other words, selected indicators need to be representative of the problem at hand and yet straightforward in order to be effectively measured and easily understood.

Characteristics of BRIDGE indicators

In the BRIDGE project, the choice of data to be used for the assessment of urban sustainability was shaped by the specific project needs of identifying trade-offs between the various planning alternatives under study. Therefore, environmental and socio-economic data was gathered for scoring different grades of sustainability of urban planning proposals. However, not all data provided by public or ad hoc statistics are available or provide relevant information.

The need to consider, in a unique and integrated assessment framework, physical processes and the respective impacts on well-being was ascertained during the interactions with planners and practitioners in the case study cities at the CoP meetings (see Chapter 12 for details). Concerns about potential

TABLE 13.2 Socio-economic policy goals identified during the case study workshops with local practitioners (CoP)

Socio-economic objectives	Athens	Firenze	Gliwice	Helsinki	London
Improve mobility	✓	✓	✓	✓	✓
Promote social inclusion	✓				✓
Ensure economic viability	✓	✓	✓		✓
Maintain public health/safety	✓		✓		
Enhance human well-being		✓		✓	✓
Optimize land use				✓	

economic and social changes were perceived as relevant for the assessment of planning alternatives. Improvement of urban metabolism processes and environmental conditions were identified as being crucial, but issues such as availability of affordable housing, reducing social exclusion and increasing employment were also articulated by city planners during the CoP meetings.

Selecting indicators thus represents an important and sensitive task which is crucial for the outcome of the decision making process. It needs to be supported by local stakeholders as well as by scientific knowledge and expertise, to ensure that they are fit for purpose and address critical and relevant issues. The adoption of participative approaches to identification and selection of indicators contributes to a focused and relevant assessment of key concerns, shared by all participants of the process.

To ensure this, the definition of indicators in BRIDGE followed a combined hierarchic bottom up and top down approach, in order to enable the identification by local stakeholders of overarching policy goals and subsequently define specific criteria for the assessment of these goals, taking into consideration data availability and modelling capacity. The indicator selection process was initiated with a bottom up approach for identifying sustainability objectives and indicators during CoP meetings in each of the case study cities. Table 13.2 illustrates the policy goals identified and agreed in each of the cities in relation to the socio-economic domain.

Socio-economic indicators in the BRIDGE case studies

Interestingly, the only common socio-economic objective across all the case study cities was the improvement of mobility. The only environmental common objective across all the cities was air quality, related to the same concern of transport and mobility. Once the sustainability objectives were identified, research was carried out on data availability at national and local levels. Identifying and examining available sustainability indicators facilitated the validation of the indicators proposed as well as the data gathering process. The overall aim of detecting appropriate and easy to access data sources resulted in an extensive list of descriptive indicators measuring standard aspects of social and economic well-being, such as GDP, employment, household income, life expectancy at birth, etc., albeit that this data was not always available at the city or neighbourhood level.

Concerning the BRIDGE case studies (see Chapter 3 for details), at the time of the project, sustainable development indicators for the Athens were only available at the national level. These are provided by the National Centre for Environment and Sustainable Development, with most up to date data published in 2003 (NCESDG 2003).

Environmental reporting in Italy developed under Agenda 21 activities is commonly undertaken by the municipalities, expanding and adapting the European common urban indicators for local sustainability

(EC 2003; Comune di Firenze 2008) according to local priorities and concerns. For Firenze, additional indicators have been developed at the regional level. The Tuscany Region has established a system for measuring sustainability at the local level with an inventory of over 1,000 indicators (CDF 2008) that can be selected and applied to the local context (e.g. suited to the urban or rural characteristics of the area). The indicators are categorized according to the sustainability objectives and their dimension:

- Economic Dimension. Objective: to create a solidary economy.
- Institutional and Social Dimension. Objective: to build a sustainable community.
- Environmental Dimension. Objective: to conserve and improve the quality of environmental resources.

Within this set, the 90 indicators that are applicable for monitoring the sustainability of Firenze are embedded (Provincia di Firenze 2005).

For Gliwice, at the time of the BRIDGE activities, no sustainability indicators were available at city level. At a national level, the Polish National Environmental Policy had started setting the measures and actions to protect natural resources and improve environmental quality and safety (CMRP 2009). Though there is an official list of sustainable development indicators in the country, it was not yet available at the time and it is mainly concerned with natural resources management and conservation. More specific policy goals connected to issues at stake in the Polish urban case study were connected to air quality, human health and noise.

For Helsinki, the Finnish national sustainable development indicators are grouped according to the strategic subject areas within the national strategy for sustainable development (FNCSD 2009), which address aspects such as the balance between use and protection of natural resources. The indicators under this strategic heading related to socio-economic issues were connected to energy consumption, use of renewables, relationships between economic growth and consumption of natural resources and energy and household expenditure for services. Table 13.3 illustrates the national sustainable indicators that are relevant to BRIDGE, for which data was available at the time, up to 2008.

TABLE 13.3 Results of the CoP meetings concerning socio-economic objectives and indicators

	Objectives	Indicators
ATHENS	Improve mobility	• Road traffic intensity; • Quality of pedestrian sideways; and • Number of parking slots.
	Maintain public health and safety	• Number and severity of road accidents and pedestrian injuries; • Number of people suffering from short term effect of air pollution (upper respiratory infections such as bronchitis and pneumonia, allergic reactions); and • Number of people suffering from long term effects of air pollution (e.g. chronic respiratory disease, lung cancer, heart disease).
	Promote social inclusion	• Extent to which roads and sideways can be used by disabled or differently able people and groups (e.g. number of safe-street-crossing points, number of repose places along the street); and • Local community composition – compared to other areas: % of elderly people, foreigners, low-income families etc.
	Promote place identity	• Aesthetic value of the area and changes due to planning intervention.
	Ensure economic viability	• Financial costs of the interventions; and • Estimated side-effects on local economy.
FIRENZE	Promote Social Comfort	• Usability of the park (number, time and type of uses); • Public appreciation of the park; • Increase/decrease on public parking spaces; and • Number of illegal activities (crime events).
	Ensure Economic Viability	• Cost associated to maintenance and pruning; and • Benefits perceived by private economic activities.
GLIWICE	Improve mobility	• Number of pedestrian streets; • Public transport use; • Length of new roads built; • Length of cycle-ways provided; and • Number of parking places built up.
	Controlled expansion of urban areas	• Number of administrative decisions; • Accessibility of district from Silesia metropolitan area (hours to/from); • Number of specific services in the district; • % of new public space; and • Increase on incomes.

HELSINKI	Cater for housing demand	• Number and type of dwellings; • Population growth; • Demand for housing types; and • Percentage of owned/rented dwellings.
	Promote social inclusion	• Access to housing; • Social class/ethnical group; • Age group of residents; and • Number of family households.
	Optimize accessibility	• Travel time to work; and • Use of public transport.
LONDON	Improve human well-being	• Number of health impacts derived "from 'heat waves' and air pollution; and • Number of residents affected by flash flooding.
	Ensure economic viability	• Cost of maintenance of green areas; • Cost of drainage; and • Value at risk of flooding.

Applying socio-economic indicators

Consultation with local experts and potential BRIDGE DSS users was undertaken to identify real-life planning proposals to be assessed using the established goals and objectives, through the identified indicators for their scoring. It is worth noting that in these CoP meetings, where BRIDGE scientists interacted with local planners, the role of qualitative indicators, including those that capture subjective valuations (e.g. aesthetic value of the area), were deemed necessary and relevant. This was particularly the case in the Firenze case study that focused on the redesign of a public green area.

The distinctive sustainability objectives for each city and the context-specific issues related to the planning alternatives to be assessed resulted in diverse assessment criteria (i.e. indicators) being proposed in each city. The results of the CoP workshops illustrate a clear link between sustainability concerns and objectives and proposed environmental indicators for each city (e.g. air quality, water and energy balance). Moreover, the commonalities in relation to planning concerns and sustainability objectives generated a consensus among the participant cities on the critical drivers of sustainable urban development, as well as facilitating an agreement on a common set of indicators applicable to all the relevant urban contexts. This consensus was achieved during the umbrella CoP meetings, where representatives from all case study cities discussed and agreed the common set of sustainability objectives and indicators. As a result of this consensus these common or core objectives and indicators were to be automatically incorporated into the DSS as standard assessment criteria (once validated with respect to the availability of data and/or model outputs). Other indicators, related to more case-specific objectives (e.g. the length of roads, associated with mobility), were made specific to particular case studies. The adopted BRIDGE methodology led to the inclusion of such indicators as discretionary indicators.

To validate the applicability of both core and discretionary indicators, a top down approach was applied by BRIDGE scientists. A key criterion for the validation of indicators for the physical processes

TABLE 13.4 Final set of socio-economic indicators

Objectives	Indicators
Common aspects (core)	
Urban land use	• New urbanized areas (land use changes)
	• Number of brownfields re-used
	• Density of development
Ensure economic viability	• Cost of intervention
	• Effects on local economy
Improve mobility and accessibility	• Quality of pedestrian sideways
	• Length of cycle ways provided
	• Length of new roads provided
	• Use of public transport
	• Number of persons close to public transport
City-specific aspects (discretionary)	
Promote social inclusion	• Access to housing and services
Maintain public health/safety	• Number of persons affected by flash flooding
Enhance human well-being	• Number of persons affected by heat waves and air pollution

consisted of the modelling capacity available within the BRIDGE project, whereas economic and social indicators were checked against the availability of existing datasets or among Geographic Information System (GIS) and modelling outputs produced in some case study sites (e.g. flash flooding affecting inhabitants in the case of Helsinki). This refinement process resulted in a reduction of socio-economic indicators to be included in the DSS due to lack of readily available socio-economic data (Table 13.4).

The final set of core and discretionary indicators is included in the DSS (see Chapter 16 for details). They are to inform decision making processes in urban planning and support sustainable development. In particular, indicators based on model outputs, simulating aspects of urban metabolism, can contribute to a more accurate prediction of potential environmental effects. Despite limitations with regards to the availability of socio-economic data, the inclusion of socio-economic considerations provides the foundations for sustainability assessment.

Conclusions

The scope of the BRIDGE project and the sustainability concerns of the case studies significantly shaped socio-economic data and indicator selection. As a result, the assessment of planning alternatives within the case studies was conditioned by a set of criteria identified through participatory approaches. This bi-directional approach to the indicator selection was paramount for ensuring that local concerns were captured and addressed during the assessment of planning proposals. The final set of indicators included environmental, social and economic considerations which can be fed with quantitative and qualitative data. Nevertheless, the incorporation and use of socio-economic indicators during the assessments was highly constrained by the lack of modelling tools for simulating socio-economic variables as well as the lack of readily available data.

References

Alberti, M. (1996). Measuring urban sustainability. *Environmental Impact Assessment Review*, 16(4–6), 381–424.

Bardos, P., Lazar, A. & Willenbrock, N. (2009). A Review of Published Sustainability Indicator Sets: How Applicable are They to Contaminated Land Indicator-Set Development? London: Cl:Aire, Contaminated Land: Applications in Real Environments. Online. Available http: <http://www.sustainableremediation.org/library/issue-papers/surf-indicator-report_1.6.09.pdf>.

Brundtland, G.H. (1987). Our Common Future. Presentation of the Report of the World Commission on Environment and Development to UNEP's 14th Governing Council Session, Nairobi, Kenya: World Commission on Environment and Development (UNEP).

Carraro, C. et al. (2013). Quantifying Sustainability: A New Approach and World Ranking. SSRN Electronic Journal. Online. Available http: <http://www.ssrn.com/abstract=2200903 (accessed March 26, 2013)>.

CDF. (2008). Linee Guida Per Un Sistema Regionale Di Indicatori Comuni Di Sostenibilità Locale. Allegato 1 – Schede Descrittive Degli Indicatori, Comune Di Firenze.

Chrysoulakis, N., Lopes, M., San José, R., Grimmond, C.S.B., Jones, M.B., Magliulo, V., Klostermann, J.E.M., Synnefa, A., Mitraka, Z., Castro, E., González, A., Vogt, R., Vesala, T., Spano, D., Pigeon, G., Freer-Smith, P., Staszewski, T., Hodges, N., Mills, G., & Cartalis, C. (2013). Sustainable Urban Metabolism as a Link Between Bio-Physical Sciences and Urban Planning: The BRIDGE Project. *Landscape and Urban Planning*, 112, 100–117.

CMRP. (2009). National Environmental Policy for 2009 – 2012 and Its 2016 Outlook. Warszawa: Council Of Ministers, Republic Of Poland. Online. Available http: <http://www.mos.gov.pl/kategoria/1979_environmental_policy/>.

Comune di Firenze. (2008). Linee Guida Per Un Sistema Regionale Di Indicatori Comuni Di Sostenibilita' Locale Allegato 1 – Schede Descrittive Degli Indicatori. Online. Available http: <http://ag21.comune.fi.it/export/sites/agenda21/materiali/linee_guida_allegato_1_def.pdf>.

DEFRA. (2009). Sustainable Development Indicators in Your Pocket; An update of the UK Government Strategy Indicators. London: Department for the Environment, Food and Rural Affairs. Online. Available http <http://www.defra.gov.uk/sustainable/government/progress/documents/SDIYP2009_a9.pdf>.

EC. (2003). European Common Indicators: Towards a Local Sustainability Profile. Report by Ambiente Italia Research Institute, Milano, Italy: European Commission. Online. Available http: <http://euronet.uwe.ac.uk/www.sustainable-cities.org/indicators/eci%20final%20report.pdf>.

EEA. (2005). *EEA Core Set of Indicators – Guide*, Kopenhagen: European Environment Agency.

EPA. (2008). Ireland's Environment. Johnstown Castle, Wexford, Ireland: Environmental Protection Agency. Online. Available http: http://www.epa.ie>.

European Commission. (1999). Sustainable Urban Development in the European Union: A Framework for Action. Communication from the Commission to the Council, the European Parliament, the Economic and Social Committee and the Committee of the Regions. Online. Available http: <http://ec.europa.eu/regional_policy/sources/docoffic/official/communic/pdf/caud/caud_en.pdf>.

FNCSD. (2009). Sustainable Development Indicators. Finish National Commission on Sustainable Development. Online. Available http <http://www.environment.fi/default.asp?node = 15131&lan= en>.

González, A. et al. (2013). A Decision-Support System for Sustainable Urban Metabolism in Europe. *Environmental Impact Assessment Review*, 38, 109–119.

Holden, M. (2006). Urban Indicators and the Integrative Ideals of Cities. *Cities*, 23(3), 170–183.

Kennedy, C., Cuddihy, J. & Engel-Yan, J. (2007). The Changing Metabolism of Cities. *Journal of Industrial Ecology*, 11(2), 43–59.

Markandya, A. & Tamborra, M. (2006). *Green Accounting in Europe: A Comparative Study*, Cheltenham, UK: Edward Elgar Publishing.

Moussiopoulos, N. et al. (2010). Environmental, Social and Economic Information Management for the Evaluation of Sustainability in Urban Areas: A System of Indicators For Thessaloniki, Greece. *Cities*, 27(5), 377–384.

NCESDG. (2003). Environmental Signals – A Report on Sustainability Indicators. Athens: National Centre for Environment and Sustainable Development, Greece. Online. Available http: <http://www.un.org/esa/agenda21/natlinfo/countr/greece/indicators.pdf>.

Newman, P.W.G. (1999). Sustainability and Cities: Extending the Metabolism Model. *Landscape and Urban Planning*, 44(4), 219–226.

Olewiler, N. (2006). Environmental Sustainability for Urban Areas: The Role of Natural Capital Indicators. *Cities*, 23(3), 184–195.

Provincia Di Firenze. (2005). Rapporto Sulla Sostenibilità – Indicatori Ambientali, Sociali Ed Economici In Provincia Di Firenze, Provincia Di Firenze – Assessorato All'ambiente.

Scerri, A. & James, P. (2010). Accounting for Sustainability: Combining Qualitative and Quantitative Research in Developing 'Indicators' of Sustainability. *International Journal of Social Research Methodology*, 13(1), 41–53.

Shen, L.-Y. et al. (2011). The Application of Urban Sustainability Indicators - A Comparison Between Various Practices. *Habitat International*, 35(1), 17–29.

Smeets, E. & Weterings, R. (1999). Environmental Indicators: Typology and Overview. Technical Report No. 25. Copenhagen: European Environment Agency.

Stiglitz, J., E., Sen, A. & Fitoussi, J.-P. (2009a). Report by the Commission on the Measurement of Economic Performance and Social Progress. Online. Available http: <http://www.stiglitz-sen-fitoussi.fr>.

Stiglitz, J.E., Sen, A. & Fitoussi, J.-P. (2009b). The Measurement of Economic Performance and Social Progress Revisited. Reflections and Overview, Paris: Centre De Recherche En Économie De Sciences Po (OFCE). Online. Available http: <http://www.ofce.sciences-po.fr/pdf/dtravail/wp2009-33.pdf>.

Valentin, A. & Spangenberg, J.H. (2000). A Guide to Community Sustainability Indicators. *Environmental Impact Assessment Review*, 20 (3), 381–392.

14

COMBINING ENVIRONMENTAL AND SOCIO-ECONOMIC DATA

Eduardo Castro and Marta Marques

UNIVERSITY OF AVEIRO

Introduction

A key objective of the BRIDGE project (Chrysoulakis et al. 2013) was to combine a set of indicators, characterizing a planning project, into a unique index reflecting the overall project contribution to the city's sustainability. Therefore, a central methodological decision to take was how to aggregate variables covering different themes and time horizons, obtained from different sources and including objective and quantitative data as well as subjective and qualitative information. A further difficulty to this challenge was the need to translate scientific data into information that can be easily understood by policy makers, without altering the rigour and significance of the message.

The more common method for assessing urban projects is Cost–Benefit Analysis (CBA), where the global evaluation of different alternatives is translated into monetary units. Nevertheless, it was concluded that CBA was not adequate for BRIDGE purposes, because the available resources and time were incompatible with the data requirements of this technique. Moreover, rather than seeking an objective valuation of projects, BRIDGE aimed to develop a heuristic tool enabling policy makers and stakeholders to assess how different preferences would influence the overall selection of project alternatives.

The solution adopted was to use a Multi-Criteria Analysis (MCA) function, where all costs and benefits considered to be relevant for the analysis are valuated according to the importance ascribed by policy makers. Rather than evaluating alternatives in absolute terms, they were measured in relation to a benchmark, which was either one of the alternatives or the option of doing nothing. This synthetic indicator must be interpreted with caution because it reflects rule of thumb and subjective valuations rather than an objective balance between costs and benefits. In synthesis, BRIDGE aimed to provide a practical tool, good to measure the sensitivity of options to preferences and to exogenous constraints. For strategic decisions, involving high investments and generating major impacts, more time-consuming and expensive evaluation techniques must be used to support final decisions.

This chapter starts with a brief comparison of CBA and MCA techniques, followed by the description of the MCA methodology developed by BRIDGE.

Combining environmental and socio-economic data for project evaluation: CBA versus MCA

The classic approach to project evaluation is to perform a CBA, where all relevant criteria are translated into costs and benefits with a monetary value (Layard & Glaister 1994). The CBA can be summarized in two steps: identification of all individuals (persons and organizations) affected by the project, followed by the quantification of all benefits and costs involved in the projects under evaluation. The benefit (or cost) of a certain impact on an individual's life can be described as the corresponding well-being change perceived by that person (Freeman 2003). It is measured as the product of the quantity of such impact by its corresponding price. Whenever it is possible, the price is defined by the market: the costs of labour, machinery and materials needed to build a house are possible examples. In many cases the impacts correspond to intangibles or externalities which are not properly valued by the market; they can only be indirectly calculated as shadow prices, reflecting the willingness either to pay for a perceived benefit or to be paid for accepting an expected negative outcome.

The quantification of the social net benefit corresponds to the sum of benefits and costs estimated for all impacts and all affected agents. The purpose of every urban project is to maximize this indicator and, as such, the planning alternative with the biggest social net benefit is chosen. This simple reasoning, applied to everyday life as well as to business decisions, is hindered by the extreme difficulty of measuring benefits and costs of a subjective nature that are felt by a heterogeneous group of individuals (Prior 1998).

Shadow prices can be estimated by econometric techniques, such as hedonic models, when the required statistical data is available. Otherwise, the solution is to use stated preferences methods, based on expensive surveys and possibly biased by respondent's cognitive limitations and unwillingness to disclose information which can affect their own interests (Whitehead 2008). Given these problems, preliminary viability studies on urban projects usually do not have the necessary time and resources to perform a proper CBA. The introduction of environmental sustainability issues creates another range of challenges. The incorporation in a CBA of restrictions imposed by environmental legislation is a difficult exercise, adding to the most complex problem of quantifying the contribution to urban sustainability of a certain project or policy. It should be stressed that a standard estimation of the shadow prices of environmental assets can be dubious, to the extent that such assets have an intrinsic value which goes beyond the interests and perceptions of stakeholders. Nevertheless, there are interesting examples of CBA applied to environmental sustainability issues, which succeed in determining the monetary value of environmental parameters such as biodiversity, air quality and esthetical importance of forests (Pearce et al. 2006; Jensen et al. 2013).

An alternative set of methodologies is being developed, by-passing the technical difficulties of CBA and allowing the simultaneous inclusion of quantitative and qualitative decision criteria. They are grouped under the general label of MCA, which encompasses 'any structured approach used to determine overall preferences among alternative options, where the options accomplish several objectives' (UNFCCC Secretariat 2008).

Any evaluation exercise begins with the identification of all the objectives of a program or a project under assessment, and with the definition of indicators that can characterize the impact on the defined objectives. In a CBA, all indicators have to be translated into monetary values. This is not the case in MCA, where techniques for comparing and ranking different outcomes are provided, allowing for the integration of different types of indicators, with diverse units and scales. Nominal, ordinal or quantitative scales can be used in the exercise of scoring, ranking and weighting each indicator.

A standard multi-criteria analysis process has three phases: the identification of the problem, the definition of the model to be implemented in its evaluation, and the evaluation of the impacts of the several options considered as a solution to the defined problem (Costa & Beinat 2011). The flexibility of this technique allows for environmental and social indicators to be developed and integrated in the same framework along with economic indicators, such as those produced in the CBA, facilitating the coexistence of several distinctive analytical dimensions. This characteristic is essential for the assessment of policy options that have monetary and nonmonetary objectives, being translated by a variety of indicators with different typologies and units.

Although the information available on the application of such techniques to the assessment of urban projects is still scarce (the development of *ex ante* assessment of urban projects is almost absent in the literature, although it is possible to collect information on *ex post* evaluation methodologies and exercises), the United Kingdom and Australia provide several examples of application of CBA and similar techniques and also a smooth transition for MCA, integrating CBA as one of the assessment components (Ruming 2006). Most examples illustrate the dependency on questionnaires to assess the valuation ascribed by residents and social agents to the expected impacts of the projects. It should be stressed again that in a preliminary assessment of sustainability there are no time and resources to implement such questionnaires, especially when the project under analysis has a small scale and a reduced impact on the city.

Simpler methodologies which can easily and quickly be implemented are needed for a preliminary evaluation of small and medium scale projects on urban sustainability. MCA can meet this challenge, since it integrates different assessment dimensions and is compatible with diverse types of data, including subjective opinions of decision makers, or stakeholders.

MCA in BRIDGE

A key goal of BRIDGE was to develop a Decision Support System (DSS) tool that performs a preliminary assessment of the impact of planning options on urban sustainability, contributing both to the definition of the best planning option and to the improvement of the chosen alternative. The selection of the evaluation method was made considering the defined goal and the restrictions faced by the project:

- there was neither enough time available nor resources to perform local surveys on stakeholder's preferences;
- the same reasons and the lack of appropriate statistical data made it impossible to apply econometric techniques to estimate shadow prices;
- the available information about the projects and their potential impact was very heterogeneous among the five case studies, creating a further incentive to build a tool with the capacity to contribute to the planning process in different stages;
- there was a considerable uncertainty concerning the description of planning alternatives impacts;
- the outputs of the physical models of BRIDGE DSS, available for all case studies, are heterogeneous in terms of meaning and capacity to characterize the impact of planning alternatives on urban sustainability, since, for each specific location, each physical parameter assumes a specific relative importance;
- in each case study there were issues requiring indicators not explicit in the BRIDGE models, which needed to be included in the assessment exercise.

Considering these constraints, the evaluation method more suited to BRIDGE was MCA, given its flexibility to adjust to different sets of inputs and its capacity to integrate information with different scales of measurements and levels of accuracy. MCA proved to be an ideal technique to rank planning alternatives, under high levels of uncertainty, lack of accurate data and the need to make preliminary assessments. The BRIDGE tool, which incorporates sophisticated models to simulate changes in the urban metabolism generated by urban projects and the consequent impacts in terms of environmental quality, is sufficiently flexible to incorporate significant upgrades in socio-economic models, if adequate data is available. This, in turn, will enable a combination of MCA and CBA techniques, creating a possible continuum where fast and cheap preliminary evaluations and rigorous analysis of large scale investments are the extreme cases.

It was also necessary to choose between the quantification in absolute or in relative terms of the contribution of each planning alternative to urban sustainability. It was decided to adopt relative values because sustainability is a too broad and complex concept to be characterized in absolute terms.

Inputs for the BRIDGE multi-criteria function

The selection of the relevant information to assess the impacts on urban sustainability is the first step in the using of BRIDGE DSS. It should be stressed again that there is not a general definition of urban sustainability and so the set of variables which should be included depend on the characteristics of the city and on the particular objectives the project aims to fulfil: for example, increasing thermal comfort in a city with hot summers, such as Athens, or using marginal land for economic and social development in the city of Gliwice. It is obvious that thermal comfort data are not relevant in the second case, though the impact on socio-economic development should always be considered in a sustainability assessment.

Therefore, the BRIDGE project faced the challenge of identifying the main areas of interest in urban projects and its relations with sustainability issues. This was made using an Analytical Hierarchy Process (AHP) approach (Saaty 1990), creating three levels of information: dimensions, criteria and indicators.

As expected, social, environmental and economic dimensions, the three pillars of sustainable development, constitute the first level of information to be gathered. Inside each dimension, the main areas of interest are selected: they are the criteria, which correspond to the second level of information. The DSS user can choose the relevant criteria for the project under analysis from a list of available possibilities: i) land use, mobility/accessibility, social inclusion and human well-being for the social dimension; ii) air quality, water balance, energy, thermal comfort as well as green spaces and materials, for the environmental dimension; iii) in the DSS, the economic dimension is not subdivided into criteria. Finally, for each criterion there is a list of available indicators. End users can add new indicators to the list, if they think they are relevant and they can find the corresponding data for all planning alternatives.

Indicators correspond to the basic information, reflecting the measurement of objective changes in the state of the world caused by the direct and indirect impacts of the project on urban sustainability. The measurement of these impacts give the indicators' scores, which need to be measured with the highest possible rigour; building and infrastructure costs, or destruction of forest area are examples of scores describing direct impacts; different components of air quality variation as a consequence of road traffic changes, thermal comfort effects of changes in buildings materials, or urban layout are scores of indirect impacts.

Each indicator has also a particular weight, reflecting its relative importance to define the criterion where it is included; weights have a subjective nature, reflecting the priorities defined by the policy makers, which in turn are expected to correspond to the preferences of stakeholders. The definition of weights is made by BRIDGE DSS end users, policy makers and technical experts. It can also be the result of participatory processes where stakeholders state their priorities and, as such, BRIDGE DSS can be a very useful instrument to improve governance procedures, by tightening the links between government and citizenship.

Definition of scores

To quantify the sustainability impacts of planning alternatives is a major challenge. BRIDGE models applied to each planning alternative generate accurate values for physical indicators of air quality, water or energy. Once these values are aggregated at the geographic (intervention area and surroundings) and temporal levels adopted by the BRIDGE models, they can easily be compared to each other. For other indicators, such as those related to the majority of socio-economic impacts, the lack of reliable data made such comparisons impossible. An available option is to ask experts to provide crude estimates of impacts relative to a reference situation, rather than absolute values.

The unavailability of absolute values for several indicators, added to the sensitivity of the function to measurement units, a problem which can only be avoided by standardization processes making scores non-dimensional, were key factors for the methodological decision of comparing planning alternatives in relative rather than absolute terms. The reference situation was either the present situation (Business As Usual (BAU) scenario) or one of the planning alternatives, whenever the comparison with the BAU scenario was not meaningful (when there were indicators which could not be defined for the initial situation). This is done by calculating the generic score S_I as the non-dimensional ratio between the indicator's values for the planning alternative J ($I_{I,J}$) and the reference situation ($I_{I,R}$):

$$S_I = \frac{I_{I,J}}{I_{I,R}}$$

For example, a score of 2 indicates that the specific planning alternative provides a value twice as big as the reference one.

Definition of weights

The methodology adopted in weights definition was a pairwise comparison of all hierarchical levels of information, that is, end users were asked to state their preference between the three dimensions, between the criteria of each dimension and between the indicators of each criterion. Saaty (2008) states that 'direct comparisons are necessary to establish measurements for intangible properties that have no scales of measurement', which is exactly the case in the BRIDGE DSS. The best way to define the relative importance of diverse indicators, referring to different and conflicting objectives, is to ask the end user very simple questions, and then translate the answers into a numerical value, reflecting the weight ascribed to that indicator when compared to the others.

This was done by presenting all pairs of combinations of the elements in each category and by defining a scale with eight levels that range from 'some importance' to 'absolutely most important'. Using the psycho-physical logarithmic Weber-Fechner response function (Saaty 2008), answers are translated into weights for all indicators, criteria and dimensions: numbers, between zero and one, adding up to one for each category.

BRIDGE multi-criteria function

Every evaluation method aims at measuring the contribution of a specific project to the social well-being, or, in order words, its social utility, considering all relevant impacts on the different agents (either positive or negative).

In generic terms, the social utility of any project can be described by:

$$U_{project} = f(S_i, \alpha_i)$$

where $U_{project}$ depends on the scores and weights of the variables (S_i, α_i), as well as on the specific functional form adopted, which in turn depends on the evaluation method selected. After calculating the social utility for each project alternative, the one with the biggest utility (positive) should be implemented. If all alternatives analysed have negative utilities or utilities similar to each other's, further studies are necessary in order to propose alternative solutions.

The simpler functional form is linear, being all the individual contributions (the products of scores S_i by the weights α_i).

$$U_{project} = \alpha_1 \cdot S_1 + \alpha_2 \cdot S_2 + \ldots \alpha_n \cdot S_n$$

This solution, however, requires the dimensional homogeneity of the indicators, something which is verified in CBA, where everything is reduced to a monetary value, but not in BRIDGE valuation exercises. To solve this problem, the solution was, as explained above, to define each variable in relation to a standard and not in absolute terms, which requires a multiplicative, rather than an additive, utility function. Moreover, it must be remembered that the weighting exercise was based on pairwise comparisons which defined the relative importance of a given dimension, criterion or indicator as an overall element and not as the unitary value of the variables. This requires that the utility function must have a unitary elasticity of substitution (Romer 2006). The Cobb–Douglas functional form

$$U_{project} = A \cdot S_1^{\alpha_1} \cdot S_2^{\alpha_2} \cdot \ldots \cdot S_n^{\alpha_n}$$

where $\sum \alpha_i = 1$ and A is a constant, fulfils both conditions, and is easy to manipulate, reasons why it was adopted by BRIDGE DSS. Concerning the function it is important to stress that each score S_i is a ratio between the impact on the variable i of the project alternative under evaluation and the impact of the BAU scenario (or the reference alternative). Each score is then a positive rational number, equal to 1 when both impacts are equal, bigger than one when the impact of the project alternative is more favourable than the reference case and smaller than one otherwise. Because the scores are relative values, the constant A must be equal to 1.

BRIDGE final assessment index

The $U_{project}$ function has a nested structure with three layers. Once scores and weights of all indicators are known, and the multi-criteria function is defined, all inputs are processed by the DSS function to produce the criteria scores. Then, since the weights of each criterion were previously defined, the DSS calculates the scores of each dimension. The combination of the dimension's scores and weights (defined by the same process), gives the multi-criteria final assessment index, which ranks the different project alternatives according to the subjective valuation of the overall set of project impacts. The sequence is as follows:

$$U_{project} = S_{Env}^{\alpha Env} \cdot S_{Eco}^{\alpha Eco} \cdot S_{Soc}^{\alpha Soc}$$

with S_i the score of the dimension i and α_i the weight of the dimension i; i is the environmental, economic or social dimension. As an example, the score of the environmental dimension is:

$$S_{Env} = S_{AQ}^{\alpha AQ} \cdot S_{WB}^{\alpha WB} \cdot S_{Soc}^{\alpha E} \cdot S_{TC}^{\alpha TC} \cdot S_{GSM}^{\alpha GSM}$$

where S_i is the score of the criterion i and α_i is the weight of the criterion i; i is the air quality (AQ), water balance (WB), energy (E), thermal comfort (TC) or green spaces and materials (GSM) criteria.

The nested structure described above (AHP) is more convenient than the use of a function where the indicator's scores and weights directly give the final index, because the valuation exercise becomes more structured and less subject to cognitive limitations. Fixing the relative importance of each dimension, then the weight of each criterion inside the respective dimension and finally the weight of each indicator facilitates a comprehensive understanding of the choices at stake and makes the selection less sensitive to the number of variables inside each category.

For example, if an unstructured multi-criteria function is used rather than an AHP structure, if it is intended to give to the environmental dimension a 50 per cent weight and if this dimension has only 10 per cent of the indicators, each weight/score combination for environmental criteria must be on average five times bigger than the equivalent values for the other dimensions. This is difficult to ensure in an unstructured multi-criteria function, making it more prone to produce all sorts of biases.

Conclusions

After a preliminary definition of the intervention's objectives, the end user is asked to select all the relevant criteria and indicators provided by BRIDGE DSS and optionally to introduce further variables and the respective data, and the pairwise valuation is then performed, producing a matrix of variables weights. These are the inputs for the multi-criteria evaluation process, based on a Cobb–Douglas function using scores and weights: the first translate the relative performance of the planning alternative under evaluation compared to a reference situation; the second reflect the relative importance ascribed by the end users to each indicator, criterion and dimension.

The DSS integrates physical model results with socio-economic data, supporting decision making by:

- promoting a more organized and systematic overview of all the possible impacts of the decision and stressing the importance of environmental issues related to urban metabolism, usually disregarded by urban planners;
- making available environmental indicators produced by complex scientific models, in a easily understandable manner;
- integrating the economic, social and environmental data into a single indicator, combining measurements of the more important project impacts with the preferences of stakeholders or policy makers;
- allowing the performance of sensitivity analyses of the overall result conditional on changes either in the weights or in the scores of indicators.

The methodology described here was not presented as the only possible answer to the challenges posed by the BRIDGE goal, but as the one more suited to comply with the restrictions faced by the DSS development. Its major advantage is the flexibility to adjust to changing assumptions concerning the projects' characteristics and to cope with different numbers of indicators characterizing each of the three sustainability dimensions.

The interactions of planning experts from all case study cities ensured that the BRIDGE DSS is able to produce useful information, easily understandable by policy makers. Nevertheless, one should keep in mind, that, as all models, the DSS output is only as good as its input data.

References

Chrysoulakis, N., Lopes, M., San José, R., Grimmond, C. S. B., Jones, M. B., Magliulo, V., Klostermann, J. E. M., Synnefa, A., Mitraka, Z., Castro, E., González, A., Vogt, R., Vesala, T., Spano, D., Pigeon, G., Freer-Smith, P., Staszewski, T., Hodges, N., Mills, G., & Cartalis, C. (2013). Sustainable urban metabolism as a link between bio-physical sciences and urban planning: the BRIDGE project. *Landscape and Urban Planning*, 112, 100–117.

Costa, C. B., & Beinat, E. (2011). Estruturação de modelos de análise multicritério de problemas de decisão pública. In J. S. Costa, T. Dentinho, & P. Nijkamp (Eds.), *Compêndio de Economia Regional, vol. II - métodos e técnicas de análise regional* (1st ed., pp. 663–697). Cascais, Portugal: Principia.

Freeman, A. (2003). *The measurement of environmental and resource values: theory and methods* (2nd ed.). Washington, D.C., USA: Resources for the Future. Retrieved from http://books.google.com/books?hl=en&lr=&id=rP9_i5lRMQ8C&oi=fnd&pg=PR7&dq=shadow+price+environemtnal+values&ots=sOATKdS2fB&sig=buNd7cqjn_ehg4VTtYJyj-VxhUw

Jensen, C., Jacobsen, B., Olsena, S. B., Dubgaarda, A., & Haslerb, B. (2013). A practical CBA-based screening procedure for identification of river basins where the costs of fulfilling the WFD requirements may be disproportionate–applied to the case of Denmark. *Journal of Environmental Economics and Policy*, 2, 164–200.

Layard, R., & Glaister, S. (1994). Introduction. In *Cost-benefit analysis* (pp. 1–56). Cambridge, UK: Cambridge University Press. doi:http://dx.doi.org/10.1017/CBO9780511521942

Pearce, D., Atkinson, G., & Mourato, S. (2006). *Cost-benefit analysis and the environment: recent developments* (p. 315). Paris, France: Organisation for Economic Co-operation and Development. Retrieved from http://trid.trb.org/view.aspx?id=795431

Prior, M. (1998). Economic valuation and environmental values. *Environmental Values*. Retrieved from http://www.ingentaconnect.com/content/whp/ev/1998/00000007/00000004/art00003

Romer, D. (2006). *Advanced macroeconomics* (3rd ed.). New York, USA: McGraw-Hill/Irwin.

Ruming, K. (2006). *MOSAIC urban renewal evaluation project: urban renewal policy, program and evaluation review* (p. 123). Kensington, Australia: City Futures Research Centre, University of New South Wales. Retrieved from https://149.171.158.103/sites/default/files/upload/pdf/cf/research/cityfuturesprojects/urbanimprovement/researchpaper4.pdf

Saaty, T. (1990). How to make a decision: the analytic hierarchy process. *European Journal of Operational Research*, 48, 9–26.

Saaty, T. (2008). Relative measurement and its generalization in decision making why pairwise comparisons are central in mathematics for the measurement of intangible factors the. *RACSAM – Revista de la Real Academia de Ciencias Exactas, Físicas y Naturales*, 102, 251–318.

UNFCCC Secretariat. (2008). *Compendium on methods and tools to evaluate impacts of, and vulnerability and adaptation to, climate change. Final draft report.* Retrieved from http://dev.cakex.org/sites/default/files/compendium on methods.pdf

Whitehead, J. (2008). Combining revealed and stated preference data to estimate the nonmarket value of ecological services: an assessment of the state of the science. *Journal of Economic Surveys*, 22, 872–908.

PART IV
The BRIDGE DSS

15

THE BRIDGE IMPACT ASSESSMENT FRAMEWORK

Ainhoa González, Mike Jones and Alison Donnelly

TRINITY COLLEGE

Introduction

The assessment of potential impacts of development proposals on the environment is a requirement under European law. The requirements for environmental impact assessment are embedded in planning legislation through Directive 2001/42/EC (CEC 2001), also known as the Strategic Environmental Assessment (SEA) Directive which requires an assessment of potential significant impacts resulting from plan (e.g. land use, transport) or programme (e.g. waste management) implementation to be conducted. In addition, Directive 2011/92/EU (CEC 2011), amending 85/337/EEC (EC 1985) and 97/11/EC (CEC 1997) – also known as the Environmental Impact Assessment (EIA) Directive – establishes the requirements for assessing potential environmental effects associated with certain specific projects (e.g. roads, housing states). In the context of urban planning, impact assessment can be defined as a systematic process of predicting and evaluating the likely effects of implementing a development proposal and thus ensure that any significant effects are appropriately addressed at the earliest appropriate decision-making stage. Such processes not only enable the early identification and mitigation of any negative effects (e.g. increased water consumption), but provide the opportunity to promote positive changes in the environment by incorporating additional environment-focused measures into the proposal (e.g. grey water capture and reuse). Although SEA and EIA focus on environmental aspects, they also address certain social aspects (e.g. human health and material assets). In practice, the assessment outcomes are considered on a par with other socio-economic aspects in the context of sustainable development. SEA and EIA procedures have similar sequential stages including:

- screening to determine the need for appraisal;
- establishing the scope of the assessment by identifying key issues to be examined and determining the geographical extent and scale of the assessment;
- establishment of the baseline environment or existing environmental conditions against which any changes incurred by the proposal can be assessed;
- identification of reasonable and realistic alternatives to achieve the plan, programme or project objectives;
- assessment of alternatives with regards to potential impacts on population, human health, biodiversity, fauna, flora, soil, water, air, climatic factors, material assets, cultural heritage, landscape and any interrelationship between the above factors;

TABLE 15.1 Key differences between SEA and EIA

	SEA	*EIA*
Objectives	Assessment of broad policies, actions	Assessment of local single project
Spatial context	Extensive geographic zoning	Spatially specific development/s
Data	Qualitative and quantitative	Mainly quantitative
Methods	Expert judgement, matrix and Geographical Information Systems approaches	As per SEA but supported by field surveys
Alternatives	Many strategic, few operational	Few strategic, many operational
Mitigation	Broad policy solutions	Practical site-specific solutions
Outputs	Often generic findings, broad mitigation	Detailed findings, site-specific mitigation

- development of mitigation measures that address any identified potential adverse impacts;
- monitoring measures to follow up the implementation of the proposal and the proposed mitigation measures, and thus identify any unforeseen adverse effects and apply remedial actions; and
- consultation involving key stakeholders during scoping, and identification of alternatives, as well as to gather public perceptions on the selected alternative and assessment findings.

However, the geographical scope and level of assessment detail of SEA and EIA differ (Table 15.1). Nevertheless, in both cases, the key purpose is to minimise the potential negative environmental impacts of a proposed action while promoting sustainability, and to inform and substantiate a final planning decision (Scott and Marsden 2003; Stoeglehner, 2004; Therivel, 2004; González, 2010).

Impact assessment findings can assist in improving the sustainability of proposals by promoting responsible management of natural resources and enhanced environmental quality and, subsequently, social well-being. This is of particular significance in the urban context, as current planning practice is challenged by the need to design more environmentally efficient urban settlements. In the context of BRIDGE (Chrysoulakis et al. 2013), consideration is given to fluxes and material exchanges within a city in light of the urban metabolism approach adopted in the project. Although the assessment of urban metabolism aspects is not a requirement in either SEA or EIA processes, its inclusion can provide valuable information on the environmental quality of urban areas. Urban metabolism is concerned with the examination of inputs, outputs and storage of energy, water, pollutants, nutrients, materials and waste, taking account of the environmental and socio-economic changes that affect or result from such metabolic balances. Detailed impact assessment of the exchanges associated with diverse physical infrastructure and building types, materials and layouts (in terms of, for example, solid waste, air and water pollutants, and water and energy fluxes) can provide significant insights and help identify sustainable urban structures.

Techniques and tools supporting impact assessment and decision-making

Impact assessment approaches commonly rely on a combination of techniques and tools applied to the various procedural stages described above. Reporting (i.e. literal description and rationalisation of pertinent considerations) represents the most widely used technique for evaluating and communicating relevant information. Tables and maps are also commonly used to illustrate or summarise certain aspects. Documenting assessment findings is always supported by assessment tools including expert judgements, matrices, multi-criteria analysis, mapping and overlays using Geographical Information Systems (GIS) and modelling (González 2010). These techniques are drawn from other areas of environmental management and land use planning and provide both quantitative and qualitative data to the assessment.

- Expert judgement. Several environmental and planning experts discuss and consider the relevant environmental issues to determine reasonable alternatives, and subsequently analyse and rank them. Although widely applied, its accountability has been questioned given its potential for bias, as different opinions and divergent interpretations of environmental risk exist among and within disciplines (Therivel 2004; Rizak & Hrudey 2005). In order to increase objectivity and ensure evidence-based judgements, field surveys (i.e. gathering of specific data on site), GIS-based spatial analysis and modelling can be used.
- Matrices. Matrix-based assessments, also known as compatibility appraisals, are widely used in impact assessment practice. Comparison of proposed alternatives against environmental objectives presented in a matrix format can facilitate interpretation and analysis (Scott & Marsden 2003; Therivel 2004). Contextual arguments derived from expert judgement are commonly provided to help rationalise the assessment outcomes. Matrices allow conflicts and trade-offs between the objectives of a proposal and environmental objectives to be easily identified, but may be subjective (Therivel 2004). Moreover, they fail to address the spatio-temporal dimensions common to environmental and urban planning issues (González 2010).
- Multi-Criteria Analysis (MCA). Although MCA is commonly used in the assessment of environmental policies (e.g. Hämäläinen & Alaja 2003; Wallenius et al. 2008), it is not as widely used in impact assessment (i.e. SEA or EIA) as other methods. This technique constitutes both a framework for structuring decision-making problems which encompass multiple decision criteria and alternatives, and a set of methods to generate/elicit and aggregate preferences regarding the performance of these alternatives. As has been also described in Chapter 13, MCA techniques enable comparison and ranking of different planning alternatives and scenarios, through the structured prioritisation (i.e. weighting or determining the relative value of importance) of a variety of environmental or sustainability criteria or indicators. Proposed alternatives are scored (based on their performance) against each weighted indicator, resulting in a ranking of proposed alternatives that reflects their sustainability or environmental feasibility. MCA acknowledges that society is composed of a diverse range of communities with different values and thus allows the fact that some issues may be more important than others to be reflected through the weighting system (Therivel 2004). In addition, MCA allows transparent comparison of alternatives and can be used with quantifiable and unquantifiable data, but it has a risk of manipulation as it can lead to very different results depending on who establishes the weighting and scoring systems.
- Mapping and spatial analysis. Mapping is a widely used technique in impact assessment (João 1998; Vanderhaegen & Muro 2005; González 2012). The geographic representation of spatial baseline data and alternatives using GIS is considered to facilitate a systematic and objective assessment of proposals, while enhancing the understanding of the distribution, patterns and linkages between proposed interventions and environmental issues within a given area (DEHLG 2004; Vanderhaegen & Muro 2005; González et al. 2011). GIS-based mapping techniques can take several approaches, including overlay and weighted-overlay operations. Overlay mapping techniques can be used to map, superimpose and spatially assess sensitive environmental areas (e.g. protected landscapes or groundwater protection areas) against urban development layouts. Weighted-overlay mapping techniques combine MCA and GIS, incorporating relative weights to each of the environmental considerations/indicators and using overlay operations (Antunes et al. 2001; González 2010). They allow the aggregation of co-occurring environmental factors and their weights, enabling the spatially specific and quantitative identification of urban areas with high vulnerability to development.

Despite the enhanced transparency and objectivity of spatial assessments, they can only be applied to urban interventions and impacts that can be mapped. Moreover, assigning 'weights' to environmental criteria or indicators involves value judgements that are concealed in the assessment.

- Modelling. GIS and other modelling tools, widely applied in research but rarely applied in impact assessment practice, aim at predicting likely future environmental and socio-economic conditions (e.g. changes in air quality, soil erosion, population density employment rates). Although no single model can cover all the range of spatial and temporal scales and processes involved in impact assessment (Therivel 2004), or indeed urban plan-making, they can be usefully adapted to simulate future scenarios (e.g. likely water or air quality changes resulting from a proposal). Despite the limitations associated with uncertainties and assumptions made, models can facilitate and promote strategic thinking by anticipating change.

All impact assessment techniques and tools have benefits and limitations. In the light of the scope of BRIDGE and given that urban land use planning is intrinsically spatial, it is considered that spatial evidence and spatial approaches need to be applied to enhance plan-making. Decision-makers at all levels are commonly required to assimilate relevant information in the form of large reports prior to any planning decision. This information load has been increased as a result of the requirement to consider environmental aspects under the SEA and EIA Directives. GIS-based techniques – with their ability to organise, analyse and display spatial information – provide a plausible alternative for relieving the information burden. Moreover, GIS have the potential to facilitate more transparent decision-making for spatial planning as results can easily be communicated and decisions can be demonstrably based on spatially specific and objective evidence (González et al. 2011). In addition, they can incorporate stakeholder perceptions and expert judgements, as well as be combined with external modelling tools to predict likely future socio-economic and/or environmental conditions, based on the characteristics of given planning alternatives (e.g. Munier et al. 2004; Stevens et al. 2007; Chen et al. 2009; Vienneau et al. 2009). In summary, GIS can contribute to (i) objective, accurate and quantifiable impact prediction and assessment; (ii) evaluating the spatial and temporal variability of impacts; (iii) predicting cumulative and large-scale effects; and ultimately, (iv) presenting all relevant information in geographic and visual formats (Antunes et al. 2001; Vanderhaegen & Muro 2005; Geneletti, 2008; González et al. 2011). This contributes to providing information in a more efficient and comprehensive manner, improving understanding and raising awareness of the spatial implications of a planning intervention (Antunes et al. 2001; Carver 2003; González, 2012).

Impact assessment framework in BRIDGE

In the context of BRIDGE, SEA and EIA principles (namely consideration and assessment of alternatives, mitigation of adverse impacts on urban metabolism and stakeholder involvement) are at the core of the impact assessment framework. The BRIDGE Decision Support System (DSS) has been developed to systematically integrate urban metabolism components into impact assessment processes with the aim of accurately quantifying the potential effects of development proposals on the urban environment and thus inform planning decisions. The impact assessment framework also incorporates decision-making principles, enabling integration of end users' (e.g. planners, stakeholders, decision-makers) perceptions of the importance of assessment criteria, or indicators. The framework was developed to ensure flexibility in terms of thematic and geographical scope, scale, assessment detail and prioritisation of assessment criteria. The objective is to provide a DSS that can be tailored to the planning hierarchy requirements (e.g. assessment detail of a land use plan versus a residential development project for a given plan area), as well as to user preferences (e.g. relative importance of air or water quality indicators).

The impact assessment framework on which the DSS is based has been developed to enable: (a) assessment of specific planning interventions at the local level (e.g. alternative layouts for a new residential area) through detailed modelling; and (b) evaluation of strategic future urban development scenarios (e.g. flood risk associated with climate change) through simulations. Local level assessments resemble EIA-type procedures where high resolution and detailed datasets are utilised when assessing the sustainability of the planning alternatives. In addition, climate, energy and economic scenarios are developed and simulated within the project to facilitate SEA-type decision-making frameworks. In both cases, alternatives/scenarios are subject to a systematic assessment of impacts. Proposed planning alternatives are assessed against sustainability objectives and associated environmental and socio-economic indicators, previously defined by stakeholders through participatory approaches (i.e. Community of Practice – CoP, see Chapter 12). Increasing/decreasing trends of indicators show improving/deteriorating environmental and socio-economic conditions, informing end users about urban metabolism changes associated with each of the considered planning alternatives, which enable them to select the most sustainable option and, where necessary, identify remedial action/s to mitigate any potential negative impacts. It must be noted that BRIDGE focuses on the interrelationship between energy, material flows and urban structure. It aims to assess how a planned urban structure may affect the exchange and transformation of energy, water and pollutants within the city. Correspondingly, it examines how such energy and material flows affect socio-economic activities and how socio-economic activities affect such flows. Therefore, the BRIDGE DSS models address the following environmental receptors in line with those included in Annex 1 of the SEA and EIA Directives (CEC 2001, 2011): water (i.e. water balance, including evapotranspiration and run-off); air and climate (i.e. air quality in terms of pollutant concentration and dispersion; as well as carbon dioxide (CO_2) emissions, carbon sinks and climate change); and material assets (i.e. energy consumption and associated heat fluxes – heat islands).

The BRIDGE DSS is based on GIS to provide a spatial dimension to the assessment. GIS are used to integrate datasets, analyse the various spatial elements (e.g. environmental and socio-economic indicators), and store and visualise the results (González et al. 2013). Spatial datasets illustrate field observations, provide key parameters to the available models, and are critical to the DSS in the assessment of planning alternatives. Therefore, the resulting model outputs and the assessment results are geographically displayed in the DSS, to help the end user to better understand the problem at hand by identifying spatial patterns and correlations.

The methodological approach to the assessment of planning alternatives in the BRIDGE DSS can be divided into a number of steps (Figure 15.1), which follow a decision-making logic, and applies an MCA approach to impact assessment (González et al. 2013). BRIDGE adopts the Analytic Hierarchy Process (AHP) MCA technique developed by Saaty (1980) and applied extensively in various branches of environmental studies (Wallenius et al. 2008). AHP provides a comprehensive and rational framework for structuring a decision problem, representing and quantifying its elements, relating those elements to an overall goal and evaluating alternative solutions. The technique entails decomposing the decision problem into a hierarchy of more easily understood sub-problems, each of which can be analysed independently. Due to the nature of the DSS, and to the specific requirements in terms of definition of objectives, criteria, indicators and weights (i.e. AHP sub-problems – Figure 15.2), the assessment is an interactive process where the end user selects from multiple choices shaping the assessment. The Helsinki case study is used for representative purposes.

Step 1: defining objectives, criteria and indicators for assessing planning alternatives

The first step requires the end user to define the sustainability objectives (e.g. improve air quality), criteria (e.g. atmospheric PM_{10} threshold/limit set by European Union (EU) legislation) and

FIGURE 15.1   Flow diagram illustrating the steps of the BRIDGE impact assessment framework

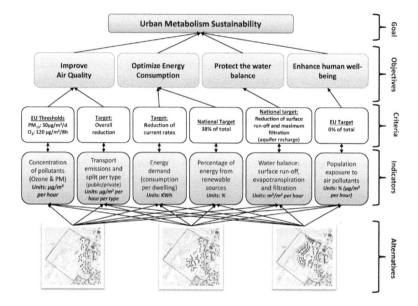

FIGURE 15.2   Example of AHP decision-making tree in BRIDGE, as applied in the Helsinki case study

indicators (e.g. atmospheric PM_{10} concentration) that are to be applied in the assessment of urban planning alternatives. This selection is commonly based on the policy objectives or planning priorities for a given urban context. The DSS contains a database with the sustainability objectives, criteria and associated 'core' (i.e. common to all case studies) and 'discretionary' (case-specific) indicators explored and agreed during the CoP meetings (Table 15.2). Although the end user can select the relevant objectives and indicators from this list, additional sustainability objectives and indicators can also be added if necessary, and assuming indicator values and data are available from the end user for input into the DSS.

End user choices in relation to sustainability objectives, criteria and indicators enable the automatic construction of the AHP hierarchy (Figure 15.2). The hierarchy consists of a decision goal and sustainability objectives, one or more criteria/indicator levels and planning alternatives for reaching such objectives.

Step 2: calculating and setting indicator values for each alternative

In this step, indicator values are provided for each alternative considered. Where the indicators can be modelled within BRIDGE, the values for the indicators selected are automatically provided to the

TABLE 15.2 BRIDGE sustainability objectives and 'core' and 'discretionary' indicators

Environmental	
Objectives	*Indicators*
Common aspects (core)	
Improve air quality	• Concentration of pollutants (PM_{10} and $PM_{2.5}$, O_3, NOx) ($\mu g/m^3$) • CO_2 and other GHG emissions (tonnes) • Number of days above established air quality threshold (days above established $\mu g/m^3$)
Improve energy efficiency	• Energy demand (kw per hour per m^2) • Potential for renewable energy (% of total energy produced) • Additional heat generated (W/m^2) • % of energy created from renewables (% of total energy produced)
Anticipate CC (flooding)	• Flooding zones (m^2) and hot spots
Optimize water use and management	• Surface runoff evapotransporation and filtration (mm^3/m^2 or m^3/m^2) • Water consumption per capita (m^3/capita)
City-specific aspects (discretionary)	
Increase green space areas	• Density of green areas (m^2 per inhabitant) • Canopy/green surface or area newly created (m^2) • Accessibility to green areas (distance in m)
Thermal comfort	• Ambient and surface air temperature (°C) • Number of days above established threshold
Optimize materials used	• Volume of material reused (days above established °C)
Socio-economic	
Objectives	*Indicators*
Common aspects (core)	
Urban land use	• New urbanized areas (land use changes in m) • Number of brownfields reused (number and m^2) • Density of development (building units/m^2)
Ensure economic viability	• Cost of intervention (€) • Effects on local economy (number of new employment opportunities)
Improve mobility and accessibility	• Quality of pedestrian sideways (qualitative) • Length of cycleways provided (m) • Length of new roads provided (m) • Use of public transport (passengers per public mode) • Number of persons close to public transport (inhabitants within 500 m)
City-specific aspects (discretionary)	
Promote social inclusion	• Access to housing and services (qualitative)
Maintain public health/safety	• Number of persons affected by flash flooding (numbers)
Enhance human well-being	• Number of persons affected by heat waves and air pollution (numbers)

TABLE 15.3 AHP scale and example of a pair-wise comparison matrix in which the three indicators concerning air quality are compared

Air quality	Concentration of pollutants	Emissions of GHG	Population exposed
Pollutant concentration	1	1/3	1/5
Emissions of GHG	3	1	1/3
Population exposed	5	3	1

1 – Equal importance
3 – Moderate importance
5 – Strong
7 – Very Strong
9 – Extreme

end user as a modelling output as shown in Plate 14. The results of the models are displayed in both spatial and averaged numerical formats. Therefore, for indicators modelled within the project, the DSS displays the spatial distribution of the indicator values within the study area in the form of a GIS map enabling the end user to contrast, for example, different building layouts within the development area and adjust them according to any identified land use conflicts or impact distribution patterns. Similarly, the value for the area is spatially averaged to obtain an overall value for that indicator and, thus, facilitate its aggregation with the non-spatial indicators.

Where models are not available within BRIDGE to compute certain specific indicators, the end user is prompted to input the relevant indicator values (as total values that may or may not be spatially specific) to progress the assessment. The multiple indicators can be combined in a spatial manner to explore the spatial distribution of their performance.

Steps 3 and 4: weighting indicators according to their priority and normalising weights

Once indicators are selected and their values defined, the end user is requested to perform a pair-wise comparison of the relevant indicators (applying AHP), enabling the integration of end user perceptions into the assessment. Each indicator is contrasted against another and the end user is prompted to determine which one is more important/significant. As a result, the indicators are weighted according to their importance/significance set by the end user and based on the sustainability goals (or planning priorities) for the city or other considerations of a subjective nature.

Using a number of pair-wise comparisons, the AHP assesses all components of a given level, and subsequently assesses the components of a lower level with respect to each component of the next-higher level (Figure 15.2). During the pair-wise comparisons, judgements are expressed in numerical values that can be processed and compared over the entire range of the problem. In the case of BRIDGE, the AHP uses an underlying scale with values from 1 to 9 to describe the relative preferences for two criteria where 9 is significantly more important than 1 (Table 15.3). The result of the pair-wise comparisons is a reciprocal quadratic matrix (Table 15.3). It allows numerical weights (w_j) to be established, where diverse elements are compared to one another.

The matrix in Table 15.3 is to be read as follows: how important is the row component (e.g. concentration of pollutants) compared to column component (e.g. population exposed to high levels of pollutants) in the context of air quality? In the example, the answer to this question indicates that population exposed to pollutants is of higher concern or has stronger importance from an air quality perspective, than pollutant concentration or GHG emissions.

Based on the results of the pair-wise comparison matrix $A \in IR^{n \times n}$ a numerical weight (w_j), or priority for each component of the hierarchy, can be determined to normalised values. This is achieved using the algorithms below:

1. Estimate the maximum eigenvalue λ_{max} of the comparison matrix, which fulfil the formula below:

$$\det\left(A - \lambda \times I\right) = 0$$

2. Determine the solution $\tilde{w}$:

$$\left(A - \lambda \times I\right) \times \tilde{w} \ = 0$$

$$\tilde{w}_i \ \geq 0$$

3. Normalise the $\tilde{w}$:

$$w_j = \frac{\tilde{w}_j}{\sum\limits_{i=1}^{n} \tilde{w}_i}$$

Applying the above algorithms, the pair-wise comparison matrix in the example in Table 15.3 results in the following weights: Population exposed = 0.657; Emissions of GHG = 0.258; and Pollutant concentration = 0.105. The numerical weights of all elements are synthesised to yield a set of overall priorities for the AHP hierarchy. The results represent each alternative's relative ability to achieve the decision goal, so they allow a straightforward consideration of the various courses of action.

The AHP approach enables consistency checks of the pair-wise comparison matrix to be performed systematically. A consistent matrix means, for example, that if the decision-maker establishes that objective x is equally important to objective y (so the comparison matrix will contain value axy = 1 = ayx), and that objective y is more important than objective w (ayw = 9; awy = 1/9); then objective x should also be more important than objective w (axw = 9; awx = 1/9). Decision-makers or end users are often not able to express consistent preferences in the case of multiple criteria. AHP measures the inconsistency of the pair-wise comparison matrix, setting a consistency threshold that cannot be exceeded and, thereby, ensuring that the weights are consistent or prompting the end user to re-establish conflicting weights.

Step 5: scoring indicator values according to their performance for each alternative

The performance of indicators is automatically defined based on how close the indicator value is to a reference basis (i.e. criteria). When assessing the performance of indicators, targets/thresholds are used as reference points to establish the nature of the indicator's performance. Thresholds refer to the maximum value permitted according to European and national legislation (i.e. upper benchmark such as the 50 $\mu g/m^3$ limit for atmospheric PM_{10}). In some instances, targets are applied referring to the minimum value the indicator should have (i.e. lower benchmark such as a minimum of 68 per cent of employment).

Where a comparable baseline exists (e.g. business as usual scenario in an urbanised area), alternatives can be contrasted against such do-nothing alternative (in the case of BRIDGE, the 2008 baseline). Where the baseline alterative is not comparable (e.g. where a forested area will be converted into an urban settlement as is the case in the Helsinki case study) the planning alternatives can be compared against one of them (e.g. the simplest one, which in the case of Helsinki includes the low density residential development) to establish which one presents the most sustainable option. A detailed description of the scoring process has been given in Chapter 14; however, it is necessary to clarify if the increase or decrease in the indicator value represents more or less sustainability (i.e. improvement or deterioration of environmental and/or socio-economic considerations). When an increase in value of an indicator means a decrease in the performance value of the alternative, the indicator score is the inverse.

Step 6: combining relative weights and values of indicators and aggregating criteria to obtain relative values for each alternative

The score of each objective is automatically calculated as a function of indicator scores and weights. Similarly, the total score for a given alternative can be calculated as a function of the total objectives scores and weights, or of the total indicators scores and weights.

The relative importance of each indicator (defined in Step 3 and normalised in Step 4) is combined with the ratios between indicators for the alternatives being compared (calculated in Step 5). The indicators are operationalised to ensure that there is no double-counting or overlap of selected indicators (e.g. energy consumption and CO_2 emissions), as well as to ensure that contrasting indicators within the same criterion (e.g. population density and number of people exposed to pollutants) are adequately addressed. These indicator values are aggregated to obtain the relative performance of one alternative when compared to another.

The total score for a given alternative is calculated as a function of the objectives scores and weights, or of the indicators scores and weights. This is done by applying the Cobb–Douglas function, as described in Chapter 14.

Step 7: presenting the results

The results are presented in a comprehensive manner, including both spatial and non-spatial information. A single overall value for each alternative is provided in the form of a performance index, based on the composite of the underlying objectives and indicators, to facilitate comparison between the alternatives. This conceals the performance of the individual indicators for each alternative and their relative significance and, therefore, individual objective and indicator results are provided separately to evaluate the variations among alternatives in relation to specific socio-economic and/or environmental considerations. This is done in the form of absolute or mean values for each sustainability indicator, criterion or objective (see Plate 3), or represented by means of a spider diagram which illustrates the performance of each indicator set (see Plate 2). Therefore, the impact assessment results combine summary values (e.g. total relative score of each alternative) with detailed values from each of the assessment stages (e.g. partial indicator values and weights or spatial distribution of indicator values) as illustrated in Plates 15 and 16. The purpose of providing such an array of formats is to reach a wide audience and to inform/support decision-making on the premise that the more information provided, the more informed the decision. Results are compiled and provided as a comprehensive summary for each of the alternatives by including:

- GIS maps available for the spatial-indicators;
- mean, maximum or minimum values for the study area for spatial indicators;
- absolute indicator values for the non-spatial indicators;

- spider diagram combining the indicator results for each objective or criterion; and
- the performance index (or total assessment value) for each alternative.

The above results are intended to facilitate the end user, planner or decision-maker in making an informed decision on the suitability of alternatives by looking at how the different alternatives affect the socio-economic and environmental components of the urban context.

Conclusions

The impact assessment model developed and applied in BRIDGE is tailored to the specific project requirements. It applies decision-making theory and impact assessment principles promoting a participative, systematic and coherent approach. The combined integration of environmental (e.g. air quality and land cover), urban metabolism (e.g. material and energy fluxes within the city) and socio-economic considerations and parameters (e.g. human well-being or cost of proposed planning alternatives) in the DSS facilitates a comprehensive and holistic approach to sustainability assessment, applicable to varied urban and planning contexts, and is demonstrated by the application of the DSS to the diverse range of case studies. Moreover, the flexibility of the DSS enables the incorporation of additional objectives and indicators, and thus associated models, their results, field measurements and/or observations if and when these become available. The system can be adapted in order to address aspects not covered within the scope of the project or to undertake impact assessment of different case studies to those considered in BRIDGE; the only requirement being input of available indicator data.

The DSS enables sensitivity analysis and promotes public participation by allowing end users to modify weights and, therefore, adjust the values to reflect a given set of priorities and/or explore the effect that public perceptions may have on selecting the most sustainable planning alternative. The array of formats for presenting the DSS results facilitates the provision of user-friendly information that is easy to understand by a wide audience. In this way, the results provide a composite scientific and social view of the relative significance of parameters and sustainability of planning interventions, providing a more holistic approach to urban planning and assessment. Moreover, it can be argued that the BRIDGE results contribute to raising awareness of significant environmental issues, particularly those associated with urban metabolism, and socio-economic considerations, and promotes their appropriate consideration within the capacity of the planning system.

References

Antunes, P., Santos, R. & Jordão, L. (2001). The Application of Geographical Information Systems to Determine Environmental Impact Significance. *Environmental Impact Assessment Review*, 21, 511–535.

Carver, S. (2003). The Future of Participatory Approaches Using Geographic Information: Developing a Research Agenda for the 21st Century. *Urban and Information Systems Association Journal*, 15, 61–71.

CEC. (1997). Directive 97/11/EC, of 3rd March, amending Directive 85/337/EEC on the Assessment of the Effects of Certain Public and Private Projects on the Environment. *Official Journal of the European Union*, L0011, 3.1997.

CEC. (2001). Directive 2001/42/EC, of 27th June, on the Assessment of the Effects of Certain Plans and Programmes on the Environment. Commission of the European Communities. *Official Journal of the European Union*, L 197/30, 21.7.2001.

CEC. (2011). Directive 2011/92/EU, of 13th December 2011, on the assessment of the effects of certain public and private projects on the environment (codification). *Official Journal of the European Union*, L 26/1, 28.1.2012.

Chen, J., Hill, A.A. & Urbano, L.D. (2009). A GIS-based Model for Urban Flood Inundation. *Journal of Hydrology*, 373(1–2) 184–192.

Chrysoulakis, N., Lopes, M., San José, R., Grimmond, C.S.B., Jones, M.B., Magliulo, V., Klostermann, J.E.M., Synnefa, A., Mitraka, Z., Castro, E., González, A., Vogt, R., Vesala, T., Spano, D., Pigeon, G., Freer-Smith, P., Staszewski, T., Hodges, N., Mills, G. & Cartalis, C. (2013). Sustainable urban metabolism as a link between bio-physical sciences and urban planning: the BRIDGE project. *Landscape and Urban Planning*, 112, 100–117.

DEHLG. (2004). Implementation of SEA Directive (2001/42/EC) Assessment of the Effects of Certain Plans and Programmes on the Environment – Guidelines of Regional Authorities and Planning Authorities. Government of Ireland: Dublin.

EC. (1985). Directive 1985/337/EEC, of 27th June, on the Assessment of the Effects of Certain Public and Private Projects on the Environment. *Official Journal of the European Union*, L0337, L175, 5.7.1985.

Geneletti, D. (2008). Impact Assessment of Proposed Ski Areas: A GIS Approach Integrating Biological, Physical and Landscape Indicators. *Environmental Impact Assessment Review*, 28, 116–130.

González, A. (2010). *Incorporating Spatial Data and GIS to Improve SEA of Landuse Plans: Opportunities and Limitations – Case Studies in the Republic of Ireland*. Germany: Lambert Academic Publishing.

González, A. (2012). GIS in Environmental Assessment: A Review of Current Issues and Future Needs. *Journal of Environmental Assessment Policy and Management*, 14, 1–23.

González, A., Donnelly, A., Jones, M., Chrysoulakis, N. and Lopes, M. (2013). A Decision-Support System for Sustainable Urban Metabolism in Europe. *Environmental Impact Assessment Review*, 38, 109–119.

González, A., Gilmer, A., Foley, R., Sweeney, J. & Fry, J. (2011). Applying Geographic Information Systems to Support Strategic Environmental Assessment: Opportunities and Limitations in the context of Irish Land-use Plans. *Environmental Impact Assessment Review*, 31, 368–381.

Hämäläinen, R.P. & Alaja, S. (2003). The Threat of Weighting Biases in Environmental Decision Analysis. Helsinki: Helsinki University of Technology. Online. Available HTTP: <http://www.e-reports.sal.hut.fi/pdf/E12.pdf>

João, E.M. (1998). Use of Geographic Information Systems in Impact Assessment. In: Porter, A. & Fittipaldi, J. (Eds.), *Environmental Methods Review: Retooling Impact Assessment for the New Century*, The Army Environmental Policy Institute (Atlanta), pp 154–163.

Munier, B., Birr-Pedersen, K. & Schou, J.S. (2004). Combined Ecological and Economic Modelling in Agricultural Landuse Scenarios. *Ecological Modelling*, 174, 5–18.

Rizak, S. & Hrudey, S.E. (2005). Interdisciplinary Comparison of Expert Risk Beliefs. *Journal of Environmental Engineering and Science*, 4, 173–185.

Saaty, T.L. (1980). *The Analytic Hierarchy Process*. McGraw Hill, New York.

Scott, P. & Marsden, P. (2003). Development of Strategic Environmental Assessment Methodologies for Plans and Programmes in Ireland (2001-DS-EEP-2/5). Environmental Protection Agency: Ireland.

Stevens, D., Dragicevic, S. & Rothley, K. (2007). iCity: A GIS–CA Modelling Tool for Urban Planning and Decision Making. *Environmental Modelling and Software*, 22, 761–773.

Stoeglehner, G. (2004). Integrating Strategic Environmental Assessment into Community Development Plans – A Case Study from Austria. European Environment, *The Journal of European Environmental Policy*, 14, 58–72.

Therivel, R. (2004). *Strategic Environmental Assessment in Action*. Earthscan, IncNet Library.

Vanderhaegen, M. & Muro, E. (2005). Contribution of a European Spatial Data Infrastructure to the Effectiveness of EIA. *Environmental Assessment Review*, 25, 123–142.

Vienneau, D., de Hoogh, K. & Briggs, D. (2009). A GIS-based Method for Modelling Air Pollution Exposures across Europe. *Science of the Total Environment*, 408, 255–266.

Wallenius, J., James, S., Dyer, P.C., Fishburn, R.E., Steuer, S. & Zionts, K.D. (2008). Multiple Criteria Decision Making, Multi-attribute Utility Theory: Recent Accomplishments and What Lies Ahead. *Management Science*, 54 (7), 1339–1340.

16

THE BRIDGE DECISION SUPPORT SYSTEM

Zina Mitraka, Manolis Diamantakis and Nektarios Chrysoulakis

FOUNDATION FOR RESEARCH AND TECHNOLOGY — HELLAS

Introduction

The increasing complexity in environmental management has emerged from the consideration of multiple environmental, social and economic factors and thus the related industry is becoming more sophisticated in the use of technology. The involvement of different groups of decision-makers generates the need for the development of reliable, comprehensive and easily accessible tools. Such tools are meant to provide information in a structured way, in a form easily understood and interpreted by the different groups of stakeholders and decision-makers in planning processes, which are fundamentally heterogenic and usually follow different interests.

A Decision Support System (DSS) is a computer-based information system intended to help people to compile useful information, identify issues, assess them and help in making decisions (Böhner 2006). A DSS is capable of supporting complex decision-making and of solving semi-structured or unstructured problems through an interface, which presents information and evaluation results in a readily understandable form. DSS are developed and used for a variety of applications, covering a wide range of thematic areas, like, biomedical applications, business applications, management and forecast etc. (Jao 2012).

Decision-making situations that involve geospatial information are quite complex, making it difficult for individuals to process all of the necessary information. Human cognitive deficiencies in memory and analysis abilities prevent decision-makers from efficiently addressing complex spatial problems or issues and the use of DSS in those cases is essential. This kind of system – also referred as Spatial DSS, or SDSS – helps describe the evolution of a phenomenon in space and time, provides knowledge-based formulation of possible actions, simulates consequences or actions of decision possibilities and assists in the formulation of implementation strategies (Sugumaran & De Groote 2011).

The basis of geospatial decision support is the Geographic Information System (GIS) technology. The basic functionality of GIS includes data management to extend human memory, graphic display to enhance visualization and spatial analysis functions to extend human computing performance. Beyond these common GIS decision aids, special features include modeling, optimization and simulation functions required to generate, evaluate and test the sensitivity of computed solutions. Other functions, such as statistical, spatial interaction and location/allocation models, can be also supported by a DSS (Nyerged 2010).

As DSS technology is moving forward, developments involving Planning Support Systems (PSS) are getting underway (Batty 2008). A PSS is a generic term for describing the variety of computer-based tools that urban and regional planners have been using for decades. The focus recently is on how to make use of decision support capabilities incorporating GIS and analytic models in a planning context. The rising pressure for ensuring sustainability in the urban environment leads planners to take into account the environmental, economic and social considerations at once and analyze the potential impacts. Therefore, there is an arising need for developing specific evaluation methods and tools to address multiple inter-disciplinary aspects within decision-making in urban planning.

DSS in sustainable urban planning: the BRIDGE system

The BRIDGE project (Chrysoulakis et al. 2013) illustrated the advantages of considering environmental issues in urban planning, by focusing on specific components of urban metabolism: energy, water, carbon and pollutant fluxes. A DSS prototype was developed (Chrysoulakis et al. 2010) as part of the BRIDGE project, introducing urban metabolism components in the planning process combined with socio-economic aspects. The BRIDGE DSS provides the potential to evaluate planning actions, which better fit the goal of changing the metabolism of urban systems towards sustainability.

A structured assessment of methods is integrated in the BRIDGE DSS, for the comparative analysis of urban planning alternatives, their ranking and selection from among them. With the use of numerical modeling and simulations, it is possible to evaluate how planning alternatives modify the fluxes of energy, water, carbon and pollutants. This is further translated as impact to the urban sustainability with the use of suitable indicators, which make it possible to combine with socio-economic indicators using a Multi-Criteria Evaluation (MCE) approach (González et al. 2013).

To cope with the complexity of urban metabolism issues, objectives are defined in relation to the interactions between the environmental elements (fluxes of energy, water, carbon and pollutants) and socio-economic components (investment costs, housing, employment, etc.) of urban sustainability. As described in detail in Chapter 15, the evaluation of the performance of each alternative is done according to the relative importance ascribed by the user to each objective. Indicators, organized in groups, are used to characterize objectives. The planning alternatives are finally ranked by assessing the performance of all indicators selected as relevant in each particular case and the evaluation outcomes depending on the objectives that the user establishes. Different objectives can be set and prioritized, some of which may conflict, and the final evaluation is a trade-off between objectives and their relative importance.

The BRIDGE DSS demonstrates the advantages of having a tool for assessing the behavior of key urban metabolism components in relation to social and economic aspects for urban planning and it highlights the necessity for further research and development in the field. The prototype and its main components are described in this chapter. The system architecture is outlined and the main components of the DSS are presented in the following sections. The method implemented to evaluate the performance of the urban planning actions using the prototype is described, and examples of the evaluation process are also presented and discussed.

System architecture

Depending on use, the architecture for building a DSS varies. Despite the differences, the operational functionalities of a DSS can be roughly subdivided into: (a) data management; (b) models, including both decision and simulation models; and (c) Graphical User Interface (GUI). Data management refers to the organization of the data into databases, easily and quickly accessible to the models and the user.

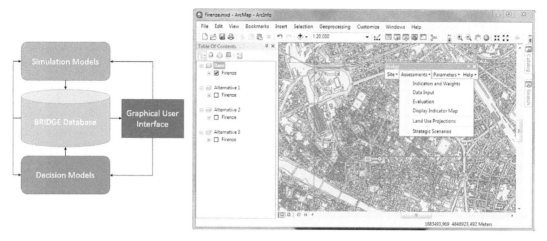

FIGURE 16.1 Simplified diagram of the BRIDGE DSS architecture (left) and the BRIDGE DSS toolbar in the ArcGIS environment (right)

In DSS terminology, by *models* we usually refer to the models used for alternatives / scenarios evaluation (*decision models*). In the case of the BRIDGE DSS, numerical models were implemented in the DSS for simulating energy, water, carbon and pollutant fluxes and they are referred as *simulation models*. The user interface is the most important component of a DSS, since it provides the link between the user (decision-maker), the data and the models. The user, not the system, is the one making the decisions, thus, it is important to ensure good understanding of the DSS processes and results. For spatial DSS, user interfaces are graphical (GUI) and they are often built based on existing platforms to exploit the respective GIS software capabilities.

A simplified diagram of the BRIDGE DSS architecture is shown in Figure 16.1 (left). The different components of the DSS are shown in this diagram, as well as the connections between them. Every arrow in the diagram represents a procedure implemented in the DSS as a module. In short, the BRIDGE database feeds the models with input data and also receives model outputs (simulations). Encoder and decoder modules are implemented for this reason. Both input and output data visualization is possible through the GUI. In addition, a dedicated module of the GUI is used for accessing the models parameters. The BRIDGE DSS prototype was built in Visual Basic (Schneider 2000). SQL (Din 1994) was used for database development and the GUI was developed as part of the ArcGIS software (Ormsby 2004) to allow the user to exploit its extended GIS capabilities for visualization and further processing at will. Figure 16.1 (right) shows the BRIDGE DSS toolbar in the ArcGIS environment. More details on each module are provided below.

Database

The BRIDGE database is the core of the system. All components of the DSS are linked to the database and the implemented procedures are accordingly developed to interact with it. Although Figure 16.1 presents the BRIDGE database as a whole, the diversity of the data for this application necessitated a more complicated geo-database structure and several processors for the data transformations and adjustments to the models' needs. Data collected from the case studies during the implementation of the project (see Chapters 5, 6 and 13 for more details on the data collection) are included in the database. This data is processed by the DSS modules and is made available to the *simulation* and *decision models* for calculations, which in turn return their outputs to the database. Special procedures (encoders and decoders) were developed to implement the interactions between the database, the models and the GUI. All datasets are easily accessible to the user through the GUI, to view, further analyze, process and update the database at will.

Simulation models

The *simulation models* are the most essential components of the BRIDGE DSS. All the information related to the environmental assessment of the planning alternatives is derived with the use of state-of-the art numerical models that simulate physical flows. Extensive details on the simulations are given in Part II. Some of the *simulation models* are very demanding in terms of computation power. This made impossible to directly integrate them in the prototype. Those models were therefore run in a powerful computer cluster for all the BRIDGE case studies (for all the respective planning alternatives) for a period of one year with an hourly time step. The simulation results were post-processed and then integrated in the DSS database.

The nature of planning alternatives was different for each of the case studies examined in BRIDGE and thus different implementations were used in the *simulation models*. The planning alternatives for each case study were interpreted to fit the specifications of the models and the system. In cases, for example, where a planning alternative implied the development of new buildings, the land use types were changed accordingly to match this intervention. In the case of the application of cool materials, different albedo values were assigned to the respective areas. In cases of planting vegetation, or green roof implementation, fractional vegetation values were modified accordingly in respect to urban surface cover.

Decision models

A *decision model* was developed in the framework of BRIDGE for the evaluation of planning alternatives taking into account the performance of the indicators described in Chapter 15. The BRIDGE *decision model* combines environmental, social and economic indicator values with the user preferences to assign appraisal scores to the planning alternatives. The user can decide on the indicators to include in the evaluation, as well as on their relative importance. A MCE algorithm is used to measure the intensity of the interactions among the different indicators and the preferences.

To overcome problems such as model uncertainties and the lack of precise socio-economic information regarding the alternatives, the adapted *decision model* compares alternatives, rather than estimating absolute appraisal scores. One possible planning alternative is considered by the user as the reference one and the other alternatives are compared to it. The reference could be the actual situation, or one of the proposed alternatives. Thus the estimated appraisal scores do not refer to absolute estimations, but rather measure the performance of each alternative compared to the reference.

Values for the environmental and socio-economic indicators are estimated and used as input to the *decision model*. Appropriate algorithms for each case were developed to aggregate the *simulation model* results, at both geographic (intervention area and surroundings) and temporal (annual) levels, resulting in values for environmental indicators. Thresholds are used in some cases to establish the nature of the indicators' performance. Information provided along with the detailed description of planning alternatives is used to estimate values for socio-economic indicators, for instance the cost of intervention or the total length of new roads.

Furthermore, as described in detail in Chapter 15, scores depicting the performance of the alterna-tives compared to the reference situation are estimated for each indicator. Scores for indicator groups are estimated in a similar manner. The overall score of each planning alternative is calculated in the same way as the groups' scores, using a function of groups' scores and weights.

In summary, the decision model adapted in BRIDGE DSS is based on a value function using scores and weights: the first one translates the relative performance of the planning alternative under evaluation when compared to a reference situation, while the second translates the relative importance of indica-tors (and indicator groups) ascribed by the user. Results are more than just one appraisal score for each alternative. The final appraisal score is simply a representation of all the collected information. The user may examine their performance by assessing individual indicator scores, as well as the scores of indicator groups. Indicator values can also be visualized, in respect to their geographical and temporal variability.

GUI

The user interface is always a very important aspect of any application developed. For the BRIDGE DSS, the GUI aimed at facilitating the set-up of the models and at illustrating the model outputs in a way easily and correctly interpreted. When dealing with spatial multi-criteria decision problems it is important to present all the components of the problem in a comprehensive and structured manner. The user is the one making the decision, thus, it is important to ensure the good understanding of the DSS processes to enable correct parameterization and to ensure good representation and explanation of the results. A balance has to be kept though, because revealing too much information about the processes in the analysis may confuse the user.

As aforementioned, the BRIDGE prototype was built in a GIS environment, which provides extended functionality on visualizing and further processing spatial data. Information associated with the case studies, describing the current situation and the planning alternatives, is included in the database and can be easily accessed using the GUI. The various datasets are presented in a self-explanatory and comprehensive way and they are categorized per planning alternative to facilitate further processing, using both the DSS capabilities and other GIS functions as well.

For the model set-up, a set of input dialog windows were developed as part of the GUI. The main concepts introduced in BRIDGE (the use of indicators to access the performance of alternatives, what specific indicators stand for, the hierarchical approach, the relative importance of indicators and indica-tor groups, etc.) are depicted in the GUI. Detailed descriptions and tutorials, as well as information on the modules are provided in the documentation accompanying the DSS.

The representation of the DSS results is a challenge, because of the complex nature of urban sustain-ability and the diversity of possible users. The DSS makes assessments of the behavior of indicators in relation to the user's preferences to answer the question, 'which alternative is better, when compared to the reference situation', on the basis of a single number. However, this number shouldn't be regarded as the only DSS output, since the DSS can do much more than that. It assesses the performance of indica-tors and indicator groups and the user has access to all this information through graphical representation to enable comparisons, as shown in Plate 15. Moreover, spatio-temporal representations are available to the user through a tool that generates different kinds of maps of indicator values, as shown in Plate 16. The maps produced are made available to the user for further processing and analysis.

Evaluate planning alternatives with the BRIDGE DSS

This section describes the procedure for evaluating planning alternatives using the BRIDGE DSS prototype. The aim is to reveal the system's functionality and potential through a step-by-step walkthrough. The procedure is represented graphically in Chapter 3 (Figure 3.1). The system provides all the necessary tools for the user to set up the evaluation according to his preferences, run several examples, choose between different types of results' visualization and post-process the results, with the ultimate goal to decide upon the optimum alternative.

The first step in the evaluation procedure is the representation of the case study and the planning alternatives in question. Upon selection of the case study, the user has access to detailed description of the planning alternatives and, more importantly, can visualize several spatial and non-spatial data featuring the current situation and the proposed planning actions.

Then, the user has to set up the parameters for evaluation according to his preferences. At first, the indicators to be included in the evaluation need to be defined. The list of available indicators is presented in a structured way in a dedicated window, shown in Figure 16.2, and the user is asked to select from among them. The hierarchical organization of indicators in three main groups depicting the environmental, social and economic dimensions of sustainability, is mapped in this structure (Figure 16.2).

The relative importance of indicators (and indicator groups) is defined by slide-bar windows for each group, as for example in Figure 16.3a. The user has to decide if he wants to change the relative importance of any indicators, or indicator groups, from equal, to more or less important. Normalized weights are computed by the system using the method described in Chapter 15. All levels of the hierarchy are accessible through this main window and relative importance can be defined at all levels of hierarchy if the user wishes to change it from the default (equal importance).

The calculation of selected indicator values for each planning alternative follows. This is performed in different ways depending on the nature of the indicator. Environmental indicator values arise from

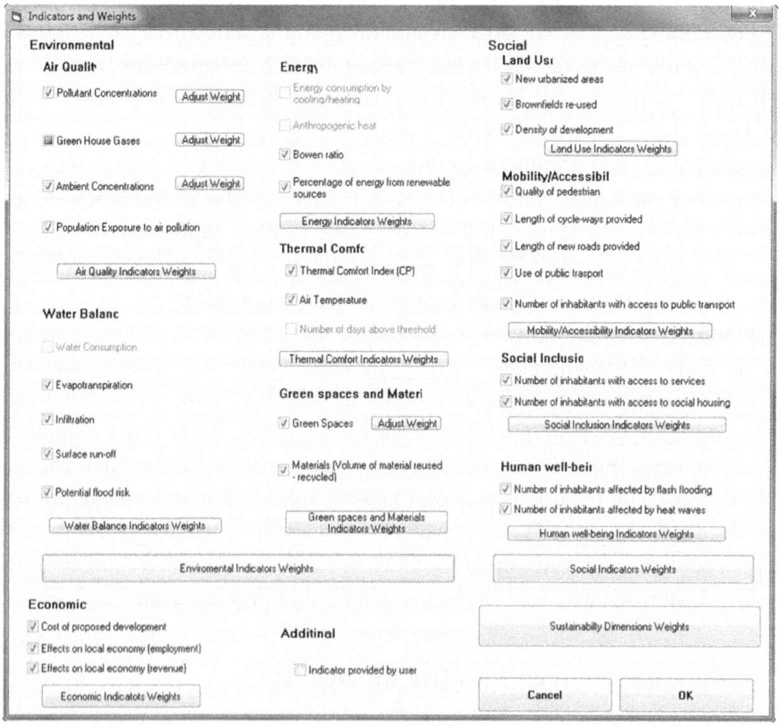

FIGURE 16.2 Selection of indicators input window of the DSS GUI

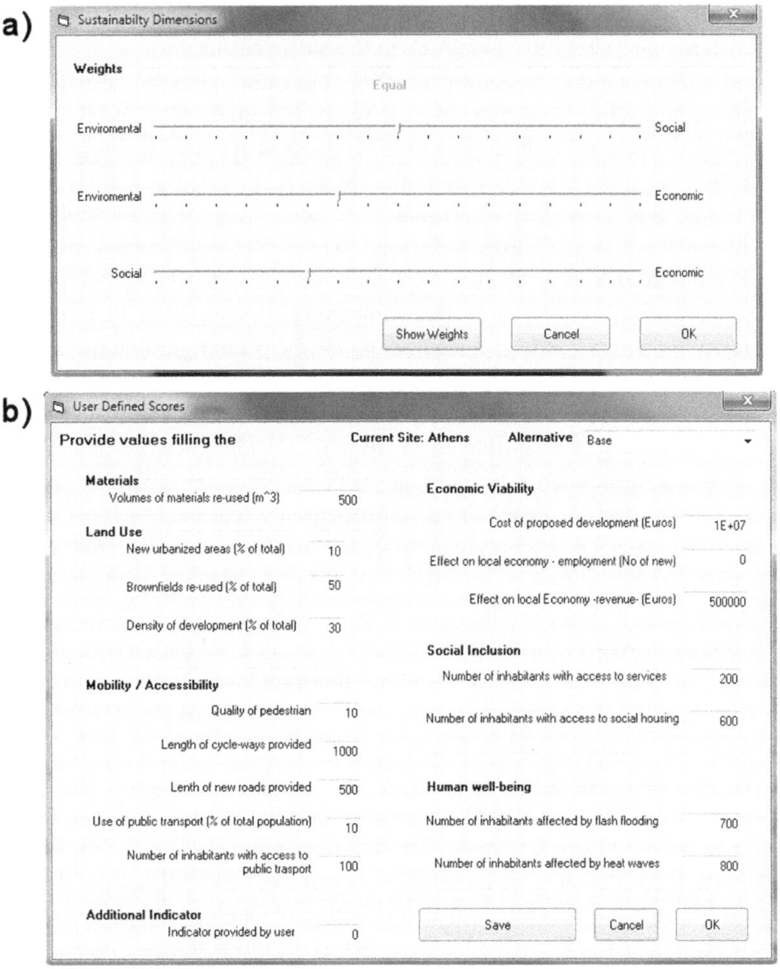

FIGURE 16.3 (a) Adjustment of relative importance input window and (b) socio-economic indicator values input window

physical flow simulations by the *simulation models*, while socio-economic indicators reflecting objective values (number of houses constructed, number of jobs created, etc.) are given as data attached to planning alternatives. As mentioned above, some *simulation models* were very demanding in terms of computational power. Thus, they run off-line and their simulation results were stored in the DSS database. However, some other models were integrated into the on-line system. The latter can be parameterized and run by the user through the DSS GUI.

No *simulation models* were used for socio-economic indicator value estimation in BRIDGE, since that was out of the scope of the project. Therefore, the user is required to provide these values for each planning alternative, using the form shown in Figure 16.3b. There is a choice of providing absolute or relative indicator values. Relative values reflect the relative performance of indicators between the alternative in question and the reference situation. Allowing for relative values is very useful, when the absolute values of indicators cannot be defined. For example, regarding the employment indicator, one may not know how many job positions a planning alternative will create, but it might be estimated that they will double compared to the reference situation. This is a good fit with the general methodology followed in BRIDGE, which assesses the performance of each planning alternative in relation to the reference situation.

Once all the parameters for the evaluation are set up, indicator scores are calculated from the *decision models* in all levels of hierarchy, as relative values between the alternatives and the reference situation. Spider diagrams are used to graphically represent the individual appraisal scores of all planning alternatives. In a spider diagram, the reference situation is always represented as a circle, having all appraisal scores equal to 1. Appraisal scores for the planning alternatives may be either higher or lower than 1, which indicate better or worse performance than the reference situation, respectively. A final appraisal score for each planning alternative is calculated as a combination of the above scores and weights, which depicts the overall performance of the alternative compared to the reference. Plate 15 shows an example of the evaluation results form. The final appraisal scores for the planning alternative are shown at the top of the window, and individual appraisal scores on the right. Each color in the spider diagram represents an alternative. The representation of all scores in one diagram facilitates comparisons. Careful interpretation of a spider diagram provides valuable information for deciding between alternatives.

This information is not enough for deciding on the best alternative. Indicator maps can reveal more information necessary for decision-making. Indicator maps are representations of the indicator values aggregated in space and time. The user is given the option to choose between different visualizations of indicator maps, aiming at providing a clear picture of the spatial and temporal distribution of the indicator values in the area. The user adjusts the desired time period and chooses between different statistics to apply for visualization. Many combinations are available, to produce seasonal maps, daily maps or yearly maps at will. The option to visualize maps of differences between the alternatives and the reference situation is also provided, to allow comparisons in accordance to the BRIDGE decision model, which is based on relative comparisons.

Additional tools are provided by the DSS, to help the user draw conclusions about the best planning alternative. These include the projection of the intervention in the land use of the broader area and in future time, given extreme future scenarios. The first is accomplished by a cellular automata model, which is integrated in the DSS and is used to simulate land use dynamics (Blecic et al. 2009). This model serves the purpose of determining future spatial distribution of city-wide land uses, taking into account the local interactions between different land uses, as well as the physical, environmental and institutional factors and other relevant characteristics of each cell. Thus, it enables accounting for the broader effects of planning decisions, in terms of a spatial distribution of land use types.

Finally, the BRIDGE DSS provides the ability to evaluate user's priorities in response to different extreme future situation scenarios. Three Strategic Scenarios were considered in the framework of the BRIDGE project, as described in detail in Chapter 17. The values of the environmental indicators in these Strategic Scenarios cases are modeled, by projecting to the year 2030 the energy, water, carbon and pollutant fluxes for the reference situation and for all the planning alternatives considered. For these projections, assumptions on environmental conditions were made, based on the Intergovernmental Panel on Climate Change (IPCC) scenarios A2, A1F1 and B1 (IPCC 2000). The socio-economic indicator values for Strategic Scenarios were defined by the Communities of Practice (see Chapter 12). The evaluation of the planning alternatives against Strategic Scenarios is done by adjusting the indicators' relative importance considering the extreme situation outlined by each scenario. The underlying decision process is similar to the basic one, adjusted for the extreme conditions of scenarios, and new appraisal scores are calculated for each Strategic Scenario for all planning alternatives. Both the cellular automata model and the evaluation of the Strategic Scenarios module provide valuable information to assess the performance and the robustness of the proposed alternatives in the future, supporting in this way the final decision.

Example of use

An example of the use of the DSS is briefly presented in this section, with selected results, aiming to demonstrate the system's functionality. The example uses the case study of Athens, Greece, and more precisely the area of the Municipality of Egaleo, where planning interventions are foreseen. Egaleo is located close to the center of Athens, with little urban vegetation, a lot of traffic and thermal discomfort issues. As discussed in Chapter 3, the proposed planning alternatives were: (a) to apply cool materials on the buildings and the roads of the Egaleo municipality; (b) to transform a brownfield area (Eleonas) to residential area; and (c) to construct a park in this brownfield area. Examples of use of the DSS to assess the performance of the proposed alternatives in relation to the 'business as usual' situation, which is regarded as reference here, are presented below.

The spider diagram in Plate 15 shows the evaluation results for Egaleo, for all indicators included in the analysis considered of equal importance. In this case, the third alternative is found to have the highest final appraisal score (0.923), compared to the reference alternative, followed by the first (0.795) and the second one (0.665). The first impression of the results is that building a park in the Eleonas brownfield is a better solution than applying cool materials on buildings and roads, or converting it to residential area. Careful interpretation of the spider diagram reveals more information. For example, the third alternative might have gained a higher overall score, but in terms of thermal comfort, the idea of applying cool materials on buildings and roads (1.12) seems a slightly better solution than creating the park (1.04). More conclusions can be drawn by adjusting the relative importance of indicators and observing the differences in appraisal scores.

Furthermore, indicator maps reveal the spatial and temporal patterns of indicator values, which are very important to account for in planning decisions. Plate 16 shows the spatial distribution of the indicator mean PM10 concentration ($\mu g/m^3$), for the summer of 2008. Higher concentrations than the surroundings are observed in all cases in the main roads. The situation does not change much between alternatives in Leoforos Athinon and Iera Odos, which are far from the area of proposed intervention (Eleonas). However, Leoforos Kifisou, which is the main street in the area Eleonas, is observed to have around a 10 $\mu g/m^3$ difference for the third alternative (Plate 16d), compared to the base case (Plate 16a), as well as to the other two alternatives (Plates 16b and 16c). Similar conclusions are extremely important for urban planning, due to the limits established by the European Union (EU). For example, according to EU Directive 2008/50/EC, the 24-hour average PM10 should not exceed 35 $\mu g/m^3$ more than 35 times in any calendar year, thus the third planning alternative is compatible with the EU Directive. More indicator maps can be drawn for different time periods and combinations with the BRIDGE DSS.

Conclusions

Several studies have addressed urban metabolism issues, but few have integrated the development of methods for the combined analysis of physical fluxes in a city environment. The methodology developed in BRIDGE is based on the application of numerical models for assessing the performance of urban planning alternatives, based on environmental and socio-economic indicators. The BRIDGE DSS integrates environmental observations with social and economic data, to evaluate planning alternatives and thus addresses and jointly examines these three pillars of urban sustainability. The DSS illustrates the advantages of using a software tool to assess the behavior of certain urban metabolism components (energy, water, carbon, air pollutants) in relation to social and economic aspects for urban planning and it highlights the necessity for further research and development in the field.

An important aspect of the evaluation process implemented in the BRIDGE DSS is the capability of accounting for user preferences. This is highly important for decision-making, since the final selection is subject to the interests and priorities of the city planners and stakeholders. The philosophy behind the method implemented in BRIDGE is to study the behavior of planning alternatives in different decision conditions and to compare the impact on sustainability for each potential implementation. In this way, it supports the user in selecting the optimum alternative with respect to environmental, social and economic criteria, concerning both the current situation and future hypotheses.

References

Batty, M. (2008). Planning Support Systems. Progress, predictions and speculations on the shape of things to come. In: Brail, R. K. (ed.) *Planning Support Systems for cities and regions*. Lincoln Institute of Land Policy, Cambridge, Massachusetts.

Blecic, A., Cecchini, G. A. & Trunfio, A. (2009). General-purpose Geosimulation Infrastructure for Spatial Decision Support. *Transaction on Computational Science VI, LNCS*, 5730, 200–218.

Böhner, C. (2006). Decision-support systems for sustainable urban planning. *International Journal of Environmental Technology and Management*, 6, 193–205.

Chrysoulakis, N., Lopes, M., San José, R., Grimmond, C. S. B., Jones, M. B., Magliulo, V., Klostermann, J. E. M., Synnefa, A., Mitraka, Z., Castro, E., González, A., Vogt, R., Vesala, T., Spano, D., Pigeon, G., Freer-Smith, P., Staszewski, T., Hodges, N., Mills, G. & Cartalis, C. (2013). Sustainable urban metabolism as a link between bio-physical sciences and urban planning: the BRIDGE project. *Landscape and Urban Planning*, 112, 100–117.

Chrysoulakis, N., Mitraka, Z., Diamantakis, E., González, A., Castro, E. A., San José, R. & Blecic, I. (2010). Accounting for urban metabolism in urban planning. The case of BRIDGE. In: CD-ROM of Proceedings of the 10th International Conference on Design & Decision Support Systems in Architecture and Urban Planning, organized by the Technical University of Einhoven, in Eindhoven, The Netherlands (July 19–22).

Din, A. I. (1994). *Structured Query Language (SQL): A Practical Introduction*. NCC Blackwell, Michigan, USA.

González, A., Donnelly, A., Jones, M., Chrysoulakis, N. & Lopes, M. (2013). A Decision-Support System for sustainable urban metabolism in Europe. *Environmental Impact Assessment Review*, 38, 109–119.

IPCC (2000). *Emissions Scenarios*. Nakicenovic, N. and Swart, R. (Eds.) Cambridge University Press, UK. pp 570.

Jao, C. (2012). *Decision Support Systems*, InTech Publications, Vukovar, Croatia. DOI: 10.5772/3371.

Nyerged, T. (2010). *Regional and Urban GIS: A Decision Support Approach*. A Division of Guilford Publications, Inc, New York, USA.

Ormsby, T. (2004). *Getting to Know ArcGIS Desktop: Basics of ArcView, ArcEditor and ArcInfo*. ESRI, Inc, Redlands, California.

Schneider, D. I. (2000). *Introduction to Visual Basic 6.0*. Prentice Hall, Virginia, USA.

Sugumaran, R. and De Groote, J. (2011). *Spatial Decision Support Systems: Principles and Practices*. CRC Press, Taylor & Francis Group, Florida, USA.

17

DECISION MAKING UNDER UNCERTAINTY

USE OF FORESIGHT FOR ASSESSING PLANNING ALTERNATIVES

Marta Marques[1], Eduardo Castro[1] and Annemarie Groot[2]

[1]UNIVERSITY OF AVEIRO AND [2]WAGENINGEN UNIVERSITY AND RESEARCH CENTRE

Introduction

Decision making is a key element of the urban planning process, involving the evaluation of different planning alternatives, based on the assessment of trade-offs between conflicting outcomes. Firstly, it requires the selection of the objectives and the identification and quantification of the impacts on each objective generated by each planning alternative. Secondly, a specific valuation must be ascribed to each objective, reflecting policy priorities, which in turn are contingent to the set of values adopted by policy makers and stakeholders and to the context provided by the exogenous state of the world. This procedure is an essential condition to warrant the quality and transparency of decision making, mainly when the interventions have long lasting effects.

However, good practice standards are often distant from routine procedures, with many decisions in urban planning either taken without the formal evaluation of alternatives or disregarding important but difficult to quantify objectives, such as those related to environmental quality and protection of natural values. Though various European directives are triggering a change in planning practice towards increasing attention paid to the evaluation of environmental impacts, there is still much scope for improvement. Rather than looking at environmental impacts as the mere fulfilment of threshold requirements, a trade-off between environmental and socio-economic costs and benefits must be adopted. The analysis of this trade-off is the main scope of the Decision Support System (DSS) that was developed in the framework of the BRIDGE project (Chrysoulakis et al. 2013). This DSS integrates models to measure different types of environmental impacts potentially generated by urban projects, as well as data to assess the economic and social impacts of the same projects.

The DSS must also be fed with data expressing the valuation of diverse objectives, which in turn can be made conditional to the materialization of different exogenous future scenarios. For example, a future world with serious economic problems and environmental stress is expected to generate different policy priorities and valuations from those arising from a more positive future.

The DSS was tested for four European cities, in a foresight workshop organized by the BRIDGE consortium. In each city, three planning alternatives were assessed for a specific urban intervention. Urban planners from the four cities and a selected group of experts were brought together to explore three extreme future scenarios and to ascribe weights to each objective in the assessment of the planning alternatives, according to each future scenario. This chapter describes the methodology used in the workshop, which combines a strategic scenario analysis with Delphi questionnaires. A short overview of the foresight methodology, with a focus on scenario analysis and Delphi questionnaires is presented in the next section, followed by the description of the foresight methodology used in BRIDGE and by the presentation of results obtained in two case studies. A brief discussion of the potential of combining these techniques with the BRIDGE DSS is also presented.

Foresight

Introduction

Urban planners, like other decision makers, face an uncertain future, driven by phenomena like economic crises, climate change and environmental degradation. Though the scope and impact of these phenomena cannot be predetermined or predicted, they must be taken into account in policies and plans. Foresight is a heuristic methodology aimed at helping decision makers to confront the uncertainties outlined above. Foresight techniques enable people to think ahead and model, create and respond to future eventualities with the aim to inform present-day decisions and mobilize joint actions (Rappert 1999).

Foresight draws on traditions of work in long-range planning and strategic planning, democratic approaches to policy analysis and participatory futures studies (Tsipouri 2002). Foresight occupies the space in which planning, policy development and future studies overlap (Gavigan et al. 2001). The use of foresight methodologies emerged in private organizations around the 1960s, as tools to analyse the potential of new technologies, and has been continuously improved and expanded to applications in a wide range of fields and, in particular, to the definition of national and regional strategies (Ahola 2003). Foresight methodologies bring together three fields of activities: futures, planning and networking (Tsipouri 2002; see Figure 17.1).

- Futures: foresight involves structured anticipation of the future as well as projections of long term social, economic and technological developments and needs, with the intent to act on this outlook. Usually futures that are at least ten years away are considered. A foresight analysis looks at multiple alternative futures and is experienced as being helpful to examine alternative paths of development, not only what is being considered to be most likely or business as usual (Kuosa 2011).

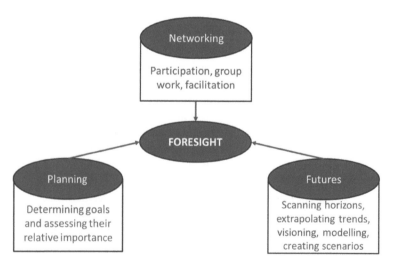

FIGURE 17.1 Process characteristics of a Foresight exercise, adapted from Miles (2002)

- Planning: foresight goes beyond regular planning, addressing operational and strategic questions. Conway (2006) emphasizes the role of foresight to integrate long term thinking into an organizational strategy development process. She distinguishes three levels in this process: i) thinking, concerned with exploration and definition of options; ii) decision making, centred on selecting and setting directions; iii) planning and implementing options. Under this logic, foresight is seen as a strategic thinking capability, occurring at the first stage of a strategy development process. By using scenarios and other exploratory methods, a foresight process generates insights into the challenges implied by prospective technological, economic and societal developments. The outcomes of a prospective analysis shape the elaboration of a strategic vision, around which there can be a shared sense of commitment, and foster present day decisions on 'what will we do' and 'how will we do it' (Conway 2006; Voros 2003).
- Networking: foresight implies that, in complex societies, knowledge relevant to longer term decision making is widely distributed, rather than centralized in government or in a few academic or corporate offices. Foresight encourages the active involvement of stakeholders with the aim of sharing knowledge and gradually building a vision of possible futures for the organization or area under stake. Even when it is not possible to establish some consensus around a vision, a foresight process should aim to promote valuable learning about possibilities and the positions of key stakeholders. Foresight methodologies use participatory methods for debating, exploring and analysing long term developments, involving a wide variety of stakeholders, often going well beyond narrow sets of experts employed in traditional futures studies or planning processes.

There is not a single method used in all foresight activities. Popper (2008) differentiates between methods that are predominantly based on: i) data processing by statistical methods and modelling; ii) opinions, reasoning and judgement by experts, brought together to interact in face-to-face or online settings; iii) creativity. In general, the choice of one particular foresight method is conditional on both the intended goals and the time and resources available for its organization. Each option has it strengths and limitations, with the combined application of multiple methods in the foresight process recommended.

In BRIDGE, the selected methodology involved the combination of a Delphi questionnaire with a scenario analysis. The methodology adopted is based on experts' knowledge and it presents the main characteristics of the foresight process:

- Futures: urban interventions have long term impacts which have to be carefully analysed; though the analysis of business as usual scenarios is a common procedure, BRIDGE considered extreme scenarios, assuming that they improve the planning process by increasing the capacity to deal with unexpected shocks.
- Planning: BRIDGE DSS is a tool that promotes good planning practices. After the selection of planning alternatives, the end users were asked to identify the more relevant indicators, according to the impacts and the goals of the intervention, and then to rank these indicators as well the three sustainability dimensions through a pair-wise comparison (see Chapter 15).
- Networking: the participants were involved from the beginning in the preparation phase; the selection of the projects to analyse was made by the Communities of Practice (CoP; see Chapter 12) promoted in all cities involved; the characterization of the projects and their socio-economic impacts were made with the CoPs close involvement; the foresight exercise itself was based on the concept of participants' interaction, as the basis to develop a general understanding of the importance of different sustainability indicators and dimensions and of how such importance varies with different scenarios.

Scenario analysis

The scenario analysis method is probably one of the most widely used methods in foresight, and it is used in decision making in the business and government spheres (Ratcliffe 2000). Scenarios consist of visions of the future and of development paths, organized in a systematic way. The term may be used to identify either a history, or an image of the future, but also a path to get there. Scenarios may be brief and descriptive, or may include story-like narratives, depending on the purpose of the analysis.

As this methodology highlights the discontinuities from the present and identifies existing options, beyond the business as usual ones, and their potential consequences, it is considered to be helpful in policy design. It also stimulates creativity and breaks the conventional obsession with present and short term problems. To be effective, scenarios must: i) fall within the limits of what might conceivably happen; ii) be internally consistent; iii) be useful to the decision making process.

Foresight exercises usually work with multiple scenarios, taking alternative courses of development into account, which demonstrate the uncertainty underlining the key questions to be answered. Scenario methods can be either exploratory or normative. The creation of explorative scenarios usually begins with an inventory and prioritization of external drivers that involve large uncertainties and highly impact the key question at stake. The more important drivers are chosen (typically three), and for each one the evolution trend, which can be positive or negative, is defined. The combinations of the different evolution trends for the chosen drivers are the bases for each scenario. Subsequently, the scenarios should be enriched with vivid and creative details, generating a history: the more absorbing, clear, compelling and entertaining the presented scenario is, the bigger the chances of its message being understood (Fahey & Randall 1998). The last phase of the exercise is the discussion of the implications of each scenario, and the analysis of strategies and actions that can be implemented to respond to the corresponding challenges. Well-designed scenarios provide a useful context for debate, leading to more robust and flexible policies and strategies, and to a shared understanding of and commitment to actions.

As pointed out by Schoemaker (2004), although scenario planning has gained many followers in industry, some academics criticized 'its subjective and heuristic nature'. The criticism is mainly based on significant misconceptions about the objective of scenario analysis and its claims. Above all, 'scenario planning is a tool for collective learning, reframing perceptions and preserving uncertainty when the latter is pervasive' (Schoemaker 2004). The usual practice among decision makers is to adopt the most plausible future scenario as the basis for policy assessment and selection, instead of considering multiple possible futures and choosing the policy option that has better results in the several possible future states.

Delphi method

In the context of future studies, the Delphi method is used for the collection, synthesis and ranking of experts' opinions, concerning emerging developments for which data is either insufficient or does not allow the simple extrapolation of trends (Gordon & Pease 2006). Delphi surveys consist of objective and clear questions on a certain thematic, with the aim of collecting information about experts' views of future developments as well as analysing the evolution of such views as an effect of the experts' interaction (Marques et al. 2009). The questions are designed to provide answers using numerical scales to measure expected outcomes or, as is the case in BRIDGE, preferences. Originally, the Delphi method was mainly conducted through postal surveys, but its application is evolving towards group meetings as well as computer and internet based methods.

Miles (2002) points out that 'the critical feature that makes Delphi different from other opinion surveys is that the survey is reiterated a number of times with the respondents receiving feedback on the responses at previous rounds (and ideally information on why judgements were made)'. Consequently, the experts' answers from one round are influenced by their colleagues' opinions given in the previous rounds. The idea behind it is that respondents can learn from interaction. The number of iterations is determined by time limitations, but ideally the end of the exercise is determined by the reaching of a consensus (Powell 2003). In the Delphi method the anonymity of the participants is desirable, namely when discussions focus on sensitive topics, or when the status or personality of some participants may dominate others (Linstone & Turoff 2002).

Delphi questionnaires and scenarios joint application

When experts' opinions depend on expectations about the future development of the external world, Delphi questionnaires can produce biased results. In addition, experts tend to disregard events that can change current trends, such as major technological developments, catastrophes or economic crises. In order to counteract such drawbacks and to increase the robustness of results, Delphi questionnaires should be combined with other foresight methodologies, such as scenario analysis (Ratcliffe 2000; Mietzner & Reger 2005; Conway 2006; Castro et al. 2012).

The evaluation of planning alternatives requires the consideration of future uncertainty, which in turn implies the designing of contrasted scenarios, determined by drivers which are not influenced by the planning options (exogenous variables). Evaluation is then made conditional on the outcomes of each scenario, supposed to influence either the project impacts or the relative valuation of each impact. Such evaluation implies expert knowledge in areas as different as economic development, environmental studies, social analysis or urban planning, which can be collected and organized through a Delphi questionnaire.

Foresight exercise: goals and methodology

Exercise goals and overall methodology

The BRIDGE DSS is a software tool designed to assess the positive and negative outcomes of planning alternatives. It combines environmental with socio-economic data and evaluates the trade-offs between different outcomes, by using a multi-criteria function, which has as inputs the scores of a given set of indicators and their respective weights. A key element of the DSS development process was the development of a foresight exercise with several objectives:

- Application of the DSS tool to BRIDGE case studies. For the DSS to be complete it is necessary to obtain weights that are not stored by the tool, because they depend on end users' choices. However, in the development phase it was essential to test how the algorithm reacts to different weights. The foresight exercise was a means to collect different sets of weights and to test the sensitivity of the multi-criteria formula. Participants took the role of end users, choosing weights.
- Assessment of how decision makers' priorities and respective indicators' weights change in response to different future scenarios, focusing on macro dimensions such as climate change, energy supply and economic performance. The combination of a scenario analysis with a Delphi questionnaire provided a good answer to such questions.
- Promoting the debate among experts and urban planning practitioners on sustainable urban policies regarding the future.
- Strengthening the interaction between the BRIDGE team and local experts and then improving the understanding of local experts' expectations concerning DSS and of the planning alternatives under evaluation.

TABLE 17.1 Dimensions' evolution in each scenario

	Climate change	Energy/technological development	Economy
BRIDGE in wonderland	+	+	+
Climate change is a burning issue	–	+	+
Lack of energy is freezing the economy	+	–	–

To achieve the above-described goals, a foresight exercise methodology was developed combining the scenario analysis method with the Delphi technique according to the seven steps below; the first three were developed by the BRIDGE team as previous work; the other ones were performed in a one day meeting between the team and the invited expert:

1. For the development of the extreme scenarios, the main drivers of urban sustainability were identified: i) climate change; ii) use of energy and technological development; and iii) economy.
2. Three extreme scenarios were created based on different combinations of the driver's evolution (positive or negative): i) BRIDGE in wonderland; ii) climate change is a burning issue; and iii) lack of energy is freezing the economy (Table 17.1);
3. The planning alternatives were studied and characterized, with the collaboration of the participants who estimated and validated the indicator's estimated scores.
4. The scenarios were presented and then discussed in thematic round tables, each focusing on one of the three sustainability components: physical (urban design); economic (urban attractiveness); and environmental (energy).
5. After a general debate on urban sustainability, participants split up again, each group dealing with a specific case study, and discussing the priorities concerning each scenario. The discussion prepared participants for the questionnaire, which was presented and handed out.
6. The Delphi questionnaire was then filled in, with the participants asked to rank the indicator's relevance (weights), according to each scenario; this ranking is the key element in selecting the planning alternatives.
7. After the presentation of the results, obtained by using a simplified version of the BRIDGE DSS, participants had the opportunity to review and eventually change their answers.

Because the quality of the expected results depends on the expertise of the participants, the invitations aimed at gathering local knowledge about the projects under evaluation by BRIDGE, taking into consideration several aspects of urban sustainability; international experts on urban sustainability were also invited, in order to ensure complementarity of interests, points of view and technical knowledge.

Design of strategic scenarios

The strategic scenarios were constructed in order to evaluate the future environmental and socio-economic impact of the land use plans or specific projects under analysis. As referred to above, though strategic scenarios are concerned with the exogenous variables that cannot be altered by local action, they affect plans' and projects' evaluations. The way drivers shape each scenario is described in Table 17.2.

Delphi questionnaire

In the Delphi questionnaire, experts were asked to define, on a scale of one to ten, the relative importance ascribed to each indicator, conditional on the three presented scenarios. This was a simplification of the

TABLE 17.2 Main drivers' configuration for each scenario

	Climate change	Energy/technological development	Economy
BRIDGE in wonderland	Low level of climate change	Gradual transition to renewable energy sources Efficient use of energy Cleaner uses of fossil energy	Socially balanced society Highly productive economy
Climate change is a burning issue	Climate change is a serious threat It is absolutely necessary: • to cut greenhouse gases (GHG) emissions • to absorb GHG already in the atmosphere	Energy is not that big a problem	Economy is growing
Lack of energy is freezing the economy	Moderate climate change	Energy shortage: • non-renewable sources are reaching the end • use of renewable sources is insufficient Reduced mobility leads to urban concentration	Resources are diverted for fast increase of renewable energy sources: • less resources for consumption • increased social inequality

DSS tool, where the ranking is obtained through a pair-wise comparison, which was necessary because of the lack of time necessary to simulate this operation. It is worth noticing that the final outcome of the pair-wise comparison is similar to the one obtained. Each indicator was attributed a weight, on a scale from zero to one, calculated as an average of the standardized values of the participants' answers.

The main difficulty in preparing the questionnaire was in the selection of adequate socio-economic indicators for each case study and in the definition of their value (see Chapter 13). The characterization of the planning alternatives of each case study was made in collaboration with local experts, from inside and outside the BRIDGE team, integrated in the corresponding Communities of Practice, described in Chapter 12. The methodology implemented in the preparation phase was the following:

- Characterization of the planning alternatives, making them operational for the DSS. A key issue was the clear understanding of the operational objectives defined by each municipality for the intervention. The physical characterization of the planning alternatives was complemented with socio-economic data.
- Identification of the relevant indicators to the evaluation. The first step of DSS application is the selection of the relevant indicators, taken from an exhaustive list provided by the tool; this task was previously performed for the foresight exercise, taking into consideration, for each case study, the objectives of the intervention and the expected impacts.
- Estimation of indicators' scores. The scores of several indicators were not rigorously defined, either because they were not yet produced by BRIDGE models at the time of the exercise, or because they corresponded to socio-economic data which could only be generated by project measurements and econometric modelling that were far beyond the scope of BRIDGE. To circumvent these limitations, BRIDGE team proposed scores for all missing parameters and the local experts adjusted them according to their more knowledgeable judgement.

- Development of the calculation process simulating the evaluation methodology used in the DSS. The evaluation process, described in detail in Chapter 14, is based on a nested methodology: the indicators are grouped into criteria and the criteria are grouped again, accordingly to the dimensions of sustainable development. In this exercise, aiming at simplifying the process by reducing the number of questions, indicators were grouped into four main criteria, in order to balance the weight of environmental and socio-economic parameters in the final result. The defined criteria, reflecting the main concerns in urban planning, were:

  o Land use and urban design: aspects related to the attractiveness of urban space (aesthetics and services supply), such as housing characteristics, equilibrium between green and built up areas, reclamation of brownfields, access to consumer services and existence of leisure infrastructures.

  o Economic viability: costs of the interventions, as well as direct and indirect benefits that they will bring to the urban economy, in particular the generation of qualified and total employment.

  o Energy balance and thermal comfort: energy is one of the main concerns nowadays, and this is also reflected in urban planning; aspects such as renewable energy production, energy consumption efficiency, outdoor thermal comfort and access to public transport (associated with energy consumption needs) are indicators associated to this criterion.

  o Physical environment: all indicators related to physical flows (air and water) and their impact on human well-being are included in this topic; examples are air quality index (synthetic index of the annual average air quality concerning nitrogen dioxide (NO_2), particulates (PM_{10}, $PM_{2.5}$) and ozone (O_3)), air quality impact on well-being, greenhouse gas (GHG) emissions, evapotranspiration, infiltration and potential flood risk.

Foresight exercise: results

The exercise provided different types of outcomes. An indirect result was to enhance the understanding and capacity of the DSS application by the BRIGE team. The debate on sustainability policies led to several suggestions and guidelines for designing urban plans and development strategies. There were different approaches in the policy debate, but the only robust policies identified were those focusing on resource efficient use and on diminishing the energy deficit in the cities.

The main results were the weights ascribed by the participants to each indicator and criteria in the three scenarios presented, which, combined with the previously defined indicator's scores, provided the key to select the best planning alternative for each scenario. This simulation helped the BRIDGE team to get a more exact notion of the final results provided by the software being developed.

As explained in detail Chapter 14, the DSS assessment method uses a Cobb–Douglas function of indicators' scores and weights. With the scores (indicators' performance) previously fixed, the Delphi questionnaire provided the weights (indicators' importance) necessary to rank the planning alternatives. This is well demonstrated in the results shown below, along with a small description of the objectives of the projects and each planning alternative. The application of the foresight exercise to the cities of Gliwice and Helsinki will be presented, as representatives of the five BRIDGE case studies described in Chapter 3.

Gliwice

In Gliwice, the construction of a new road increased the accessibility of the area where the Polytechnic University is located, as well as its capacity to attract new infrastructure and new projects. The areas that will be improved are presently occupied by a small sports zone, old industrial premises and some green spaces. The three planning alternatives for the Gliwice case study have been described in detail in Chapter 3.

The main goal is a rehabilitation project, aimed at enhancing the benefits of the new road. Complementary goals are: i) creation of an innovative economic structure; ii) improvement of quality of life; iii) development of metropolitan functions (in the context of the 14 cities which make the Silesian Agglomeration); iv) reinforcement of the public space attractiveness; and v) empowerment of civil society.

The application of the Delphi questionnaire results to the Cobb–Douglas function show that the third alternative is the most robust option: whatever the scenario defining the context where the political decision is taken, the best alternative is the construction of the sports and technological centres (Table 17.3). However, the interpretation of the second scenario results should be treated with some caution: in a situation of severe climate change, experts hesitated between the sports centre or the technological centre or both, because neither investment is related to the key interest driven by the scenario. It is also important to stress that the first alternative was used as the reference case, implying that in all cases it was ranked with the index 1 (see Chapter 14); the same applies to the other case study presented.

After a first round, the participants were invited to change their answers in order to find out if either the presentation of results or the debate that followed had any impact on their opinions. Table 17.3 shows that the change in the final outcome was small. Table 17.4 indicates the standard deviations of answers; the first row refers to weights ascribed to the four dimensions as a whole, while the remaining rows show standard deviations for weights inside each dimension. Results show that, in general, the experts did not improve their consensus on dimensions' weights, with indicators related to energy/thermal comfort the only case where standard deviations significantly decreased.

TABLE 17.3 DSS scores concerning the Gliwice case study, on the 1st and 2nd round

	Planning alternative 1		Planning alternative 2		Planning alternative 3	
Round	1st	2nd	1st	2nd	1st	2nd
Scenario I	1,00	1,00	1,04	1,01	**1,29**	**1,33**
Scenario II	1,00	1,00	1,03	1,02	**1,04**	**1,05**
Scenario III	1,00	1,00	1,07	1,08	**1,38**	**1,37**

TABLE 17.4 Average standard deviation of the questionnaires' answers, grouped by dimensions, on the 1st and 2nd round, concerning the Gliwice case study

	Scenario I		Scenario II		Scenario III	
Round	1st	2nd	1st	2nd	1st	2nd
Dimensions	0,094	0,083	0,059	0,059	0,055	0,057
Physical environment	0,079	0,074	0,075	0,072	0,078	0,078
Energy/thermal comfort	0,145	0,145	0,070	**0,058**	0,214	**0,051**
Land use/urban design	0,041	0,056	0,062	0,061	0,073	0,066
Economic viability	0,100	**0,078**	0,113	0,109	0,031	0,031

Note: bold stands for decreases of variability of more than 20%, whereas underscored stands for increases of variability of more than 20%.

Helsinki

The Finish case study focused on the construction of a residential neighbourhood in a green area of Helsinki city, located in the Meri-Rastila suburb. The three planning alternatives for the Helsinki case study have been described in detail in Chapter 3. The main goal is to increase the urban density within walking distance of the Rastila metro station (600 m radius) by creating new housing and workplaces, balancing the provision of green and built areas. Complementary goals are: i) to minimize traffic based energy consumption and carbon dioxide emissions; ii) to develop a more balanced community and dwelling stock by building more owned dwellings and bigger apartments; iii) to maintain or increase services in Meri-Rastila; and iv) to maintain a sufficient amount of green area and facilities for outdoor recreation.

The planning alternatives for Meri-Rastila presented little contrast in performances in the three scenarios (Table 17.5). When there are no economic or environmental constraints, the results for the three alternatives are almost equal: there are no clear gains of increasing the constructed area. In the second and third scenarios, the results point to the project with more inhabitants and built up area (planning alternative 3), but with a marginal advantage over planning alternative 2. The second round of answers did not bring significant changes to the average results, although the standard deviations showed contradictory patterns of convergence and divergence arising from further discussion: there was a greater consensus concerning the importance of the overall dimensions but not inside each dimension (Table 17.6). This interesting outcome indicates that broad dimensions are a much clearer basis for discussing and forging a collective opinion than detailed indicators, which require a sounder technical background.

TABLE 17.5 DSS scores concerning the Helsinki case study, on the 1st and 2nd round

	Planning alternative 1		Planning alternative 2		Planning alternative 3	
Round	1st	2nd	1st	2nd	1st	2nd
Scenario I	**1,00**	**1,00**	**1,00**	0,99	0,99	0,98
Scenario II	1,00	1,00	1,08	1,10	**1,10**	**1,12**
Scenario III	1,00	1,00	1,14	1,14	**1,16**	**1,15**

TABLE 17.6 Average standard deviation of the questionnaires' answers, grouped by dimensions, on the 1st and 2nd round, concerning the Helsinki case study

	Scenario I		Scenario II		Scenario III	
Round	1st	2nd	1st	2nd	1st	2nd
Dimensions	0,060	0,064	0,081	**0,046**	0,084	**0,055**
Physical environment	0,054	0,057	0,045	<u>0,081</u>	0,096	**0,049**
Energy/thermal comfort	0,192	0,192	0,078	<u>0,098</u>	0,133	<u>0,160</u>
Land use/urban design	0,065	0,067	0,058	<u>0,085</u>	0,076	<u>0,092</u>
Economic viability	0,065	0,066	0,072	0,073	0,026	**0,016**

Note: bold stands for decreases of variability of more than 20%, whereas underscored stands for increases of variability of more than 20%.

Conclusions

The foresight exercise was organized in order to collect a coherent set of weights for the case studies and to test the assessment methodology. However, as BRIDGE was a research project aimed at reinforcing the dialogue between practitioners and academia, it was also intended to promote the diffusion of good practices: strategic planning, envisioning the future not only as the most likely scenario, but creating conditions for the city to respond to unexpected shocks, and to incorporate technical knowledge in decision making. It is believed that the foresight exercise supplied some building blocks to this bridge.

When a municipality defines planning alternatives, this kind of exercise is very fruitful: it is important to foresee the exogenous developments that will determine the city in the future and to understand how the choices of today will affect the goals of the future. In the present case, it was emphasized that for serious energy, environmental or climate change constraints, all investments will have to be readjusted; otherwise the focus of urban policy would be the prevention of all other types of problems and the increase of quality of life, in general. Regarding the thematic round tables, some robust issues were identified:

- efficient use of resources is essential; energy is the key resource but not the only one;
- planning must be flexible, both in terms of adopted solutions and decision making processes;
- green spaces play a very important role in urban areas;
- cities' attractiveness is increasingly related to diversity in technical competences and promotion of creativity;
- urban governance requires a BRIDGE attitude, linking different technical and scientific areas, as well as linking decision making to multi-criteria tools which combine objective and technical information with scientific knowledge and subjective preferences.

The questionnaires showed the potential of the BRIDGE DSS and the way it combines environmental with socio-economic data, subject to different political preferences. It was possible to verify, as expected, that different scenarios lead to different outcomes, ensuring the DSS sensitivity to changes.

The decision making process should always involve both the identification of alternatives and their evaluation accordingly to different future developments. After reflecting on all future scenarios, politicians should balance the results and choose one alternative; when the choice is based on solid information and is the same for all alternative contexts, we can be sure that we have a good project for the city. Otherwise, when the choice varies with scenarios, or there are no clear results, the decision to be taken requires more information about the advantages and drawbacks of each alternative (additional indicators, more accurate values for indicators, rethinking of weights) and about expected future developments. For Gliwice, there was a robust alternative that was the best in all scenarios. Conversely, the preferred alternatives in Helsinki depended on future scenarios. In summary, three different cases were found:

- robust alternatives, which present the best score in all situations;
- unclear evaluation of alternatives, where the scores are very similar, indicating the need to use more and better information;
- unstable results, where the best alternative varies according to the scenarios, which reflects the need to deepen knowledge about future evolution, before a decision is taken.

It was also possible to see that most participants were not familiar with the evaluation process, which, in some ways, reduced their ability to understand the results presented and, consequently, the potential of the second round of questionnaires. The results of the sensitivity analysis showed that it is a good idea to regroup the indicators in more well-balanced dimensions, instead of having several environmental criteria, which would naturally reduce the capacity of the DSS to respond to changes in socio-economic variables and artificially increase the environmental component of the political decision. It is also possible to conclude, given the difficulty of the participants to link the questionnaires with the indicators weights, that the pair-wise comparison is a good strategy to overcome this problem.

References

Ahola, E. (2003). Technology Foresight within the Finnish Innovation System. *The Third Generation Foresight and Prioritization in Science and Technology Policy*. http://www.nistep.go.jp/IC/ic030227/pdf/p3-4.pdf

Castro, E., Marques, J. & Borges, M. (2012). Foresight methodologies. Application to the housing market. 52nd Congress of the European Regional Science Association, 21–25 August 2012 Bratislava, Slovakia.

Chrysoulakis, N., Lopes, M., San José, R., Grimmond, C.S.B., Jones, M.B., Magliulo, V., Klostermann, J.E.M., Synnefa, A., Mitraka, Z., Castro, E., González, A., Vogt, R., Vesala, T., Spano, D., Pigeon, G., Freer-Smith, P., Staszewski, T., Hodges, N., Mills, G. & Cartalis, C. (2013). Sustainable urban metabolism as a link between bio-physical sciences and urban planning: the BRIDGE project. *Landscape and Urban Planning*, 112, 100–117.

Conway, M. (2006). An Overview of Foresight Methodologies. Sydney, Thinking Futures. http://thinkingfutures.net/wp-content/uploads/2010/10/An-Overview-of-Foresight-Methodologies1.pdf

Fahey, L. & Randall, R. (1998). *Learning from the Future*, New York: John Wiley & Sons.

Gavigan, J., Scapolo, F., Keenan, M., Miles, I., Farhi, F., Lecoq, D., Capriati, M. & Di Bartolomeo, T. (2001). FOREN (Foresight for Regional Development Network). A Practical Guide to Regional Foresight. European Commission. Report EUR 20128 EN © European Communities, Brussels. http://forera.jrc.ec.europa.eu/documents/eur20128en.pdf

Gordon, T. & Pease, A. (2006). RT Delphi: An efficient. *Technological Forecasting and Social Change*, 73, 321–333.

Kuosa, T. (2011). Practicing Strategic Foresight in Government: Cases of Finland, Singapore and the European Union. RSIS Monograph No. 19. S. Rajaratnam School of International studies of Nanyang Technological University, Singapore: Booksmith.

Linstone, H. & Turoff, M. (2002). The Delphi Method. Techniques and Application. http://is.njit.edu/pubs/delphibook/delphibook.pdf

Marques, J., Castro, E. A., Martins, J., Marques, M., Esteves, C. & Simão, R. (2009). Exercício de Prospectiva para a Região Centro – Análise de Cenários e Questionário Delphi. *Revista Portuguesa de Estudos Regionais*, 19, 111–131.

Mietzner, D. & Reger, G. (2005). Advantages and disadvantages of scenario approaches for strategic foresight. *International Journal of Technology Intelligence and Planning*, 1, 220–239.

Miles, I. (2002). *Appraisal of alternative methods and procedures for producing Regional Foresight*. Report prepared by CRIC for the European Commission's DG Research funded STRATA – ETAN Expert Group Action. Manchester.

Popper, R. (2008). How are foresight methods selected? *Foresight*, 10, 62–89.

Powell, C. (2003). The Delphi technique: myths and realities. *Journal of Advanced Nursing*, 41, 376–382.

Rappert, B. (1999). Rationalising the future? Foresight in science and technology policy co-ordination. *Futures*, 31, 527–545.

Ratcliffe, J. (2000). Scenario building: a suitable method for strategic property planning? *Property Management*, 18, 127–144.

Schoemaker, P.J. (2004). Forecasting and scenario planning: the challenges of uncertainty and complexity. In Koehler, D.J. & Harvey, N. (Ed) *Blackwell Handbook of Judgment and Decision Making*. Malden, MA: Blackwell Publishing Ltd 274.

Tsipouri, L. (2002). Regional Foresight in the Cohesion Countries: Experiences, concepts and lessons learned. Brussels: European Commision - Research DG.

Voros, J. (2003). A generic foresight process framework. *Foresight*, 5, 10–21.

18

GUIDELINES FOR URBAN SUSTAINABLE DEVELOPMENT

Myriam Lopes, Carlos Borrego, Helena Martins and Jorge Humberto Amorim

UNIVERSITY OF AVEIRO

Sustainable urban planning strategies

The future of urban planning has been the object of continuous debate throughout the entire process of formulation of sustainability and climate change related strategies. To some extent, urban planning had been considered an inefficient tool, unable to address development effectively (UN-HABITAT 2010). The definition of sustainability objectives and the issues related to climate change impacts has renewed the attention towards tools that are able to influence the form of urban development and the shape of urban areas. Planning is now affirming itself as a tool to tackle not only the problems faced by urban areas, but also to devise how urban areas can positively contribute to sustainability issues in a new globalized world under the context of global climate change and global economic crises.

Factors shaping 21st-century cities include: i) the environmental challenge of climate change and cities' dependence on fossil fuels; ii) the demographic challenge of rapid urbanization, the shrinking, ageing population in some countries, as well as the increasing multicultural composition of cities; iii) the economic challenges linked to the uncertainty of future economic growth; iv) the challenges linked to social and spatial inequality, urban sprawl and unplanned urbanization; and v) the challenges and opportunities of increasing democratization, and awareness of social and economic rights in developing countries.

Urban planning, when adequately used, is a powerful tool to engage communities, shape visions and guide towards them. Agenda 21 Principles for Action has inspired urban planners and the Local Agenda 21 (LA21) movement through its motto 'act local and think global'. Local authorities construct, operate and maintain economic, social and environmental infrastructure, oversee planning processes, establish local environmental policies and regulations, and assist in implementing national and sub-national environmental policies. Since the level of governance is closest to the people, they play a vital role in educating, mobilizing and responding to the public to promote sustainable development.

According to UK Local Agenda 21 Steering Group, cited by Selman (1996), there are six key elements of a LA21:

1. Managing and improving the local authority's own environmental performance through corporate commitment, staff training and awareness raising, implementation of environmental management systems and environmental budgeting, and policy integration across sectors.

2. Integrating sustainable development aims into the local authority's policies and activities through land use planning, transport policies and programmes, tourism and visitor strategies, health strategies, and welfare, equal opportunities and poverty strategies.
3. Awareness raising and education, supporting environmental education, promoting awareness-raising events, supporting voluntary groups, publication of local information and press releases, and promoting initiatives to encourage behaviour change and practical action.
4. Consulting and involving the general public through public consultation processes, forums, focus groups and feedback mechanisms.
5. Partnerships promotion through meetings, workshops and conferences, working and advisory groups, rounds tables, developing partnerships and support.
6. Measuring, monitoring and reporting on progress towards sustainability through environmental monitoring, local state of the environment reporting, sustainability indicators and targets, Environmental Impact Assessment (EIA), and Strategic Environmental Assessment (SEA).

For its unique characteristics – being the only movement of its kind mobilizing local and regional actors around the fulfilment of European Union (EU) objectives, climate and energy policy – the Covenant of Mayors has been portrayed by European institutions as an exceptional model of multi-level governance. Signatories to the Covenant of Mayors represent cities committed to implementing sustainable energy policies to meet and exceed the EU 20 per cent carbon dioxide (CO_2) reduction objective through increased energy efficiency and development of renewable energy sources. Beyond energy savings, the results of signatories' actions are manifold: creation of skilled and stable jobs, not subject to delocalization; healthier environment and quality of life; enhanced economic competitiveness; and greater energy independence.

Impact assessment approaches, including SEA, Sustainability Appraisals (SA) and Sustainability Impact Assessment (SIA), also promote the sustainability of land use planning, by enabling the identification and mitigation of negative impacts arising from the implementation of urban development plans or specific projects (CEC 2001).

The effects of urbanization and climate change are converging in dangerous ways that seriously threaten the world's environmental, economic and social stability (UN-HABITAT 2011). The main sources of greenhouse gas (GHG) emissions from urban areas are related to the consumption of fossil fuels: energy supply for electricity generation (mainly from coal, gas and oil); transportation; energy use in commercial and residential buildings for lighting, cooking, space heating and cooling; industrial production; and waste.

Evidence is mounting that climate change presents unique challenges for urban areas and their growing populations (IPCC 2007). Although it is widely recognized that traditional urban planning models and approaches have contributed to the present environmental situation, it is also clear that addressing environmental issues at city level will not be possible without appropriate urban planning systems that integrate respect for the natural environment with the improvement of the human environment and reduction of GHG emissions of cities. The pursuit of carbon-neutral cities, through the increase in green areas within cities and the use of ecological services, often combined with efforts to close the energy/materials loops of cities, decentralize renewal energy production and reuse waste, has recently emerged as a way to increase synergies between build and natural environments in cities. Patterns of urban growth can be influenced to reduce carbon emission, by promoting more compact cities, and ensuring better access to collective or non-motorized transport. Resilience to potential environmental disasters can be increased through proper urban and land use planning.

Guidelines on sustainable urban planning

Sustainable urban development aims to achieve a healthy and high quality of life for the present and the coming generations, reducing the impact on the global and local environments, while promoting social cohesion and economic development in a way that assures satisfying living conditions for present and future inhabitants.

The contribution made by the BRIDGE project (Chrysoulakis et al. 2013) for the formalization of sustainability objectives and for providing practical tools for reaching them is based on the concept of 'urban metabolism'. In this framework, sustainable urban planning strategies seek an optimal fit between the system and its environment through the creation of a long-term direction (or vision), goals and strategies for the allocation of resources and monitoring impacts, and detailed action plans.

Urban planning is, in this context, conceived as the platform for integrating spatial, social, economic, cultural and environmental concerns in cities. The sustainable urban planning approach is achieved only through the integration of the diverse complex components of urban metabolism – energy, water, carbon and pollutants – in an overall broad strategic plan. However, due to the multifaceted nature of spatial planning, different guides do focus on different aspects of planning and tend to overlook or lay little emphasis on other aspects.

Also, the sustainable planning guidance should be a general framework that could be applied (with little or no modification) in different local contexts (Alshuwaikhat 2003). As main challenges for sustainable planning guidance nowadays it is worth mentioning:

- The urban planning process should be inclusive and participative, enhancing the economic potential of cities using urban planning as the integrative tool across sectors.
- Supporting the adoption of European and national legislation, strategies and policies for the implementation of sustainable urban design strategies, including frameworks for sustainable land use and transport planning and sustainable development plans for cities
- Promoting knowledge exchange and good practice guidance, through manuals on good practice in urban and regional governance, measures to improve public participation, promotion of environmental and integrated planning and urban design tools and methods; and improvement of cooperation between local government, the business community and private citizens on sustainable development.
- Raising awareness and promoting education, information and research in urban design for sustainability and sustainable urban development
- Enhancing the contribution of urban planning to the provision of basic services, transport and infrastructure, with a view at ensuring efficiency and equity, as well as environmental sustainability.

The following general intervention guidelines should be taken into account when developing a sustainable urban planning programme:

- Ensuring the strategic location of new developments in relation to the natural environment and transport systems.
- Promoting mixed land use to make best use of the benefits of proximity (easy and equitable access to services, amenities, green areas and workplaces) and ensure the maximum efficiency in the use of public infrastructure and services.
- Promoting sufficient density and intensity of activity and use so that services such as public transport are viable and efficient whilst achieving a high quality living environment.

- Including an adequate green structure to optimize the ecological quality of the urban areas including microclimate and air pollution.
- Promoting a high quality and well-planned public infrastructure including public transport services, pedestrian and cycle networks and networks of streets and public spaces to promote accessibility and to support a high level of social, cultural and economic activity.
- Making use of the state-of-the-art of resource saving technology including low energy housing and other buildings, environmental technology, fuel efficient, non-polluting transportations systems, recycling systems, district heating and bio-mass fuelled and other alternative forms of power production.
- Addressing climate change through an integrated and multidimensional approach, developing mitigation and adaptation measures within an urban sustainability frame, and at the same time assist in achieving less carbon-intensive patterns of urban growth.
- 'Building the city inwards', reusing existing land for new developments, most of them focused in an area around and within the inner city, connected by efficient public transport services.

Threats and barriers to be overcome include:

- lack of political will and awareness or lack of capacity for considering social, economic and environment as different aspects of the same problem;
- present administrative systems, legislation and procedures designed for sectored rather than integrated approaches;
- lack of appropriate training and education;
- the complexity of the holistic vision of sustainable development allied to the lack of knowledge sharing and interaction between different knowledge areas.

Water balance

Issues such as water balance and human consumption can be strongly influenced by urban planning. The water required for drinking and other domestic purposes is a significant proportion of the total water demand. On average in Europe 100–200 litres of drinking water per person per day are consumed. This water is distributed through a network of subsurface pipes that are often old and leaky, thus losing a significant amount. For instance, Thames Water estimates the amount of water losses to be about 217 litres per property per day. On average, the distribution water losses in Europe are about 25 per cent. Measures to reduce water loss can reduce the total water consumption significantly.

Today's changing environment poses enormous challenges at different temporal and spatial scales. Urban areas emerge as an important subject of research as the combined effect of global climate change and urban growth will likely result in an increase of urban population vulnerability to environmental problems such as heat waves, extreme precipitation or poor air quality.

Other important challenges of urban water management is to keep the city safe from flooding and safe from pests, thus providing conditions for economic prosperity. More sustainability can be achieved if more precipitation is stored instead of immediately discharged. This requires small scale measures to be implemented on a large area, such as:

- Green roofs – this roof type is vegetated with grasses and mosses; rainwater falling on the roof is absorbed first by the plant and soil cover, only after saturation of the soil and vegetation cover is rainfall transformed to direct runoff. The total reduction of runoff can range between 25–80 per cent of the total rainfall (Mentens et al. 2006). Next to the runoff reduction the absorbed radiation is also reduced in summer and the insulation of the house in winter is improved (Rahola et al. 2009), which leads to an overall reduction in energy consumption.

- Gravel roofs – gravel stones can easily be applied to flat roofs and reduce runoff on average of 25 per cent of the total rainfall (Mentens et al. 2006). The roof storage water may be infiltrated into the soil locally, thus recharging the groundwater.
- Permeable streets – have permeable cover so that water can infiltrate easily. This reduces direct runoff and stores water that will be partly available during dry periods. Moreover, efficient early warning systems, and, to a certain extent, land use planning which avoids flood prone areas and building codes taking into account impacts from floods (making buildings more flood resistant), will help to reduce damages and losses caused by these unavoidable events.

Over the next decades the urban environment will be subject to pressures resulting from climatic changes, changes in urban structure and changes in human behaviour. The Urban Heat Island (UHI) phenomenon illustrates this interaction as it may be accentuated by climate change and exacerbates the impacts of heat waves (IPCC 2007). Higher temperatures experienced under climate change, combined with UHI effect, will have impacts for air quality and human comfort. Measurements made throughout the BRIDGE case studies have shown that high densities increase temperatures on urban surfaces and prevent them from cooling down during night-time, whereas green areas and water bodies in urban areas are able to mitigate UHI effects, alongside a proper choice of surface materials and morphologies which facilitate urban ventilation.

Energy and CO_2 fluxes

When trying to reduce energy consumption, a fundamental step is to increase the energy efficiency of the urban structure. Against the background of this guiding principle a wide range of measures are available. They are as diverse as the different cities are, e.g. depending on climate and geographical location, but the general goal is a reduction in energy consumption. These measures include:

- The improvement of the infrastructure (from local, to neighbourhood, to city scale) through building insulation, installation of new windows, insulation of pipelines of central heating systems, establishing building guidelines for new settlements (from low to zero emission buildings), improving efficiency of street lighting, initiating decentralized energy networks, improving public transport and reducing distance between living and working.
- Increasing the share of renewable energy, both in energy consumption and production. Again there are many measures to be taken, spanning from fostering the use of solar thermal and/or photovoltaic energy (covering roofs or facades) from a single roof to larger solar power plants.

A by-product of optimizing the energy efficiency of the urban structure is automatically a reduction in CO_2 emissions reducing the urban share in GHG increase.

The urban energy balance, between the surface and the atmosphere, can be influenced in order to improve individual thermal comfort. A substantial measure is to change the partitioning of the available energy so that less energy is heating the atmosphere via the sensible heat flux. Most efficient is planting trees, followed by green roofs and establishing lawns and lower vegetation. Even a single tree has a positive effect on the thermal comfort on the local scale and alley trees improve the radiation climate for the pedestrians on the street canyon scale.

Another possibility is to alter the albedo (by applying paint or different cool materials) at least on a neighbourhood scale reducing the amount of energy heating the atmosphere and the energy stored in the building material and thus reducing the nocturnal UHI effect.

Air quality and human health

Current urban development patterns generate high traffic levels that have significant impacts on the environment and on the health of urban citizens, as well as on the overall quality of life in towns. Nearly all (97 per cent) of Europe's urban citizens are exposed to air pollution levels that exceed EU quality objectives for particulates, 44 per cent for ozone and 14 per cent for nitrogen dioxide (COM, 2004). Substantial progress has been made in tackling emissions from individual motor vehicles; however, hotspots continue to be a problem and the growing overall levels of motorized transport in urban areas are to some extent offsetting this progress. The high level of motorized urban transport also contributes to increasingly sedentary lifestyles with a range of negative effects on health and life expectancy, notably in relation to cardiovascular disease (McCann and Ewing 2003).

Guidelines for improving air quality based on urban planning strategies include:

- encouraging urban patterns and forms, integrating land use and transport policies that reduce the need to travel, especially by car;
- implementing traffic management measures, as an emission control measure;
- promoting public transport including vehicle fleet renovation, improved management of routes and schedules, and reallocated road space (bus lanes), among others;
- introducing schemes to encourage people to use more sustainable forms of transport such as cycle hire schemes and the installation of new cycle parking spaces;
- aiding the penetration of cleaner vehicle propulsion technologies, such as the installation of electric vehicle charging points;
- creating of Low Emission Zones with road user charging and parking charging;
- creating green urban areas to improve air quality (Beckett et al., 1998).

Finally, debates on the urban structure have become strongly polarized between the advocates and opponents of the compact and of the dispersed or sprawled city. Several modelling studies seem to confirm the dilemma, well summarized by Cervero (2001): 'exposure levels and thus health risks, are lower with sprawl, but tailpipe emissions and fossil-fuel consumption are greatly increased'.

Socio-economic issues

In addition to the environmental dimension referred to previously, sustainable development also encompasses social and economic dimensions. Urban areas should provide adequate living conditions for all inhabitants, including aspects such as employment, education, wealth, culture, social inclusion, accessibility and social security. They must also guarantee the necessary resources for urban development: human and economic. Cities' capacities to attract people and firms will, in the long term, determine which cities can maintain their development in a sustainable manner. It is necessary to trade off between different objectives that compete for scarce economic resources, with the ultimate goal of improving urban quality of life. The big issue is how to choose between different objectives, in order to promote sustainable development. There is no straightforward answer to this, since the concept has different meanings depending on the city to which it is being applied. In a certain location, it might be more important to create employment opportunities, and in another the best option can be to relocate an industrial firm, removing it from the urban area and improving air quality. Nevertheless, for one to aim at sustainable development, certain basic conditions must exist concerning housing, public services (transport, wealth, education) and accessibility. Social cohesion is a concern that must be reflected in every thematic or global policy – namely education, access to services and provision and layout of residential areas, aiming also at the prevention of ghetto creation. Furthermore, the issue of financing interventions to improve urban environmental conditions will require the development and implementation of new tools and strategies for the involvement of private capital, joining public and private efforts to improve urban quality of life and also cities' competitiveness.

Once the overall definition of sustainable development objectives is agreed, it is necessary to create a methodology to help sustainable decision making, at an urban project's level. The urban metabolism approach is a way to describe these complex interrelations, making possible the construction of a Decision Support System (DSS) that helps to clarify the trade-offs implicit in each decision. The evaluation for any urban project must be based on the functions that it will perform. The goals of the intervention must be clearly defined at the beginning of the project, not only in what concerns the main focus of the project but also on its potential impacts in other areas (secondary goals). If a green area is going to be constructed, leisure capability and impact on landscaping must be considered, along with the environmental issues. The development of a new housing area raises more complex questions concerning its functions – the new built space must: i) be attractive, but functional; ii) have as many inhabitants as possible, but not neglect local quality of life; iii) provide accessibility; iv) have basic services; and v) provide employment (conditional to the project). This list can be extended, according to each specific project.

The main problem is how to compare the benefits and drawbacks of each alternative, considering the broad scope of the aspects that urban projects present, most of them conditional on future scenarios (e.g. the value added by an industrial park depends on the future economic environment), and the intangible aspects (e.g. aesthetics quality and value of diminishing air quality problems), for which accounting methodologies are complex and subjective. Even with the problems that the methodology faces, net benefit is a sound indicator for decision making support, which has to be estimated for each option being considered in a certain area.

The process of net benefit estimation should include sensitivity analysis to indicators' values (namely when they are estimated and cannot be accurately calculated) and to the relative weighs attributed to different objectives (implicit prices). The outcome of this process is the identification of the robust alternative, which is the alternative that has the best performance, independently of the future scenarios considered.

Guidelines for evaluation of urban projects include:

- Make a thorough evaluation of each planning alternative of urban projects impacts, with public participation.
- Use a multi-criteria analysis methodology, as objective as possible, based on – but not limited to – all quantified data available.
- Avoid the different approaches usually implemented to evaluate environmental projects and socio-economic projects: methodologies should be combined in order to access the socio-economic impacts of the environmental projects, and also the impacts on urban metabolism of the socio-economic projects.
- Develop an evaluation culture, using the urban metabolism approach, integrating the different sustainability dimensions that result in a combined analysis of an urban project, capable of contributing to long-term sustainable urban decisions.

Learning with BRIDGE case studies

The contribution of case studies to the BRIDGE learning process and to the development of guidelines is illustrated here using the City of Helsinki. The aim of Helsinki is to manage the structural shifts in spatial terms to the existing city structure without eating into the green areas, without creating massive traffic congestion, and without letting house prices soar. Most of the new development is concentrated in brownfield areas (areas of degradation or industrial land surplus to requirements), thereby keeping the use of greenfield sites to an almost a negligible level. The city pays attention to social exclusion and, generally, achieves a 50 per cent minimum of social housing in each major scheme. The new centres are meant to achieve a better balance within the city structure whilst at the same time not undermine the importance of the downtown area.

There are various forest/park areas with different characters and natural features to be taken into consideration. There is lot of opposition to planning new development within existing green areas. In addition, there is a regional recreation route along the coast in Meri-Rastila (see Chapter 3 for a detailed description of planning alternatives in Helsinki case study). The plan must maintain the valuable forest areas, as well as a geological formation (an ice-age rock outcrop in the middle of the plot) on the hilltop which gives character to this forest part of Meri-Rastila.

To assess the sustainability of City of Helsinki a peer review was conducted by evaluating the differences or deficiencies which occur between the actual activities of Helsinki and the ideal model. In terms of the sustainability of traffic, it was commented that even though Helsinki has excellent public transportation, a traffic policy which integrates environmental viewpoints should also include much more. The challenges presented by regional traffic require regional cooperation which is based on political decision making in particular. Planned new cycle paths to promote cycling are promising, but in order to make cycling a viable transport alternative along with public transport and driving, financial incentives and better advertising are needed, for example.

Helsinki's air quality problems are not as considerable as in many other European cities and the city's air protection programme is of a very high quality. When it comes to noise issues, Helsinki implements its legislative obligations through its noise prevention plan, but it was evaluated as inadequate.

In energy saving, the energy efficient aims set for new buildings are a good start, but they were considered modest compared with other European cities. Helsinki should improve the energy efficiency of all the buildings in the city – not just those belonging to the municipality. Helsinki was also considered passive as a developer of renewable energy sources when compared with many European cities. To reach the climate and energy goals, Helsinki needs partnerships outside the city organization. The lack of coordination between different people and organizations was regarded as a threat to achieving these aims. The energy efficiency of building and renovation as well as decentralized energy production solutions is, above all, part of developing the city. Energy policy should be seen as an integral part of the activities of the whole city.

Helsinki is known for good quality drinking water as well as for efficient sewage treatment. The recycling of rainwater and grey water, as well as the natural handling procedures, are a few examples of water management.

From the city planning perspective the DSS developed within the BRIDGE project is offering a new alternative to quantify the impact of different city designs on the relevant aspects of urban metabolism and improve the sustainability of the urban texture. The important features in which DSS can serve the City of Helsinki are reflected in the targets/indicators identified in the Community of Practice meetings in Helsinki (see Chapter 12 for details):

- optimize energy consumption (energy demand, energy balance in buildings, share of renewable energy);

- protect water resources (water balance, water contamination, water infrastructures, percentage of imperviousness);
- improve air quality (pollutant concentration, GHG emissions, emissions from private and public transport);
- enhance human well-being (inhabitant density, population exposure to air pollutants, population exposure to polluted waters, population exposure to contaminated soils, existing urban and natural structures to promote and improve citizen health);
- anticipate climate change (carbon sinks, material reuse, number of zero-carbon buildings).

The BRIDGE DSS provided a structured presentation of planning alternatives and the tools to evaluate them on the basis of environmental impacts of energy, water, carbon and air pollutant fluxes. The quantitative estimations performed for the City of Helsinki allowed better comparison of the development alternatives, but it also set very high demands on the models and data behind these estimations.

References

Alshuwaikhat, H. M. (2003). Developing spatial urban planning guidance for achieving sustainable urban development. *Global Business and Economics Review (GBER)*, 5(3), 51–66.

Beckett, K. P., Freer-Smith, P. H. & Taylor, G. (1998). Urban woodlands: Their role in reducing the effects of particulate pollution. *Environmental Pollution*, 99 (3), 347–360.

CEC (2001). Directive 2001/42/EC, of 27th June, on the Assessment of the Effects of Certain Plans and Programmes on the Environment, Luxemburg. *Official Journal of the European Union*, L 197/30, 21.7.2001.

Cervero, R. (2001). Transport and land use: Key issues in metropolitan planning and smart growth, *Australian Planner*, 38(1), 29–37.

Chrysoulakis, N., Lopes, M., San José, R., Grimmond, C. S. B., Jones, M. B., Magliulo, V., Klostermann, J. E. M., Synnefa, A., Mitraka, Z., Castro, E., González, A., Vogt, R., Vesala, T., Spano, D., Pigeon, G., Freer-Smith, P., Staszewski, T., Hodges, N., Mills, G. & Cartalis, C. (2013). Sustainable urban metabolism as a link between bio-physical sciences and urban planning: The BRIDGE project. *Landscape and Urban Planning*, 112, 100–117.

COM (2004) 60 final, Communication from the Commission to the council, the European Parliament, the European Economic and Social Committee and the Committee of the Regions, 'Towards a thematic strategy on the urban environment', Brussels, Official Journal C 98, 23.04.2004.

IPCC (2007). Climate Change 2007: Synthesis Report. Contribution of Working Groups I, II and III to the Fourth Assessment Report of the Intergovernmental Panel on Climate Change [Core Writing Team: Pachauri, R. K and Reisinger, A. (eds.)]. IPCC, Geneva, Switzerland, 104 pp.

Mentens, J., Raes, D. & Hermy, M. (2006). Green roofs as a tool for solving the rainwater runoff problem in the urbanized 21st century? *Landscape and Urban Planning*, 77, 217–226.

McCann, B. A. and Ewing R. (2003). *Measuring the Health Effects of Sprawl*. Washington, DC: Smart Growth America, 40 pp.

Rahola, B. S., van Oppen, P. & Mulder, K. (2009). 'Heat in the City' – an inventory of knowledge and knowledge deficiencies regarding heat stress in Dutch cities and options for its mitigation, Klimaat voor Ruimte. Report number: KvR 013/2009, Delft.

Selman, P. (1996). *Local Sustainability: Managing and Planning Ecologically Sound Places*. London: Environmental Studies, SAGE, 175 pp.

UN-HABITAT (2010). State of the World's Cities 2010/2011, bridging the urban divide. United Nations Human Settlements Programme. London: Earthscan. Available online: http://www.unhabitat.org/pmss/listItemDetails.aspx?publicationID=2917

UN-HABITAT (2011). Cities and climate change: global report on human settlements. United Nations Human Settlements Programme. London: Earthscan. Available online: http://www.unhabitat.org/pmss/listItemDetails.aspx?publicationID=3085

PART V
Conclusions

19

CONCLUSIONS

Nektarios Chrysoulakis[1], Eduardo Castro[2], Eddy J. Moors[3]

[1]FOUNDATION FOR RESEARCH AND TECHNOLOGY – HELLAS, [2]UNIVERSITY OF AVEIRO AND [3]ALTERRA - WAGENINGEN UR

The urban metabolism approach by BRIDGE

This book aimed at presenting the work and the impacts of the Framework Programme 7 (FP7) funded project BRIDGE, focusing on its main outcomes: the definition and assessment of urban metabolism based on energy, water, carbon and pollutant fluxes and the development of a spatial Decision Support System (DSS) for sustainable urban planning, which takes account of urban metabolism.

Chapter 1 gave a broad introduction to urban metabolism, and the urban metabolism from micrometeorological and urban climate perspective was presented in Chapter 4. The framework of decision support tools in urban planning was described in Chapter 2. As was discussed in detail in Chapter 3, most urban metabolism studies to date use coarse, or highly aggregated data (i.e. top-down approach), often at the city or regional level, that provide a snapshot of resource or energy use, and that can't be correlated with specific locations, activities or people. The inputs and outputs of food, water, energy and pollutants have been studied across multiple cities and at the scale of the individual city. An alternative approach is the 'disaggregated approach' (i.e. bottom-up approach), which involves detailed data (or initially disaggregated data) being used; for example scaling up from individual properties to a neighbourhood. By relating the spatially explicit flows with the relevant census data and human activities, the inputs and the associated outputs generated can be assessed. Significant progress has been made in tracking energy and material flows at the building scale. The challenge ahead is to design sustainable neighbourhoods and cities by directly influencing their urban metabolism processes.

As presented in Chapter 3, the main objective of the BRIDGE bottom-up approach was to provide a structured assessment of urban metabolism processes (restricted to energy, water, carbon and air pollutants) in different planning alternatives, as well as to provide methods for comparative analysis, ranking and selection, in support of urban planning decisions. Chapter 3 also described the BRIDGE case studies, located in five selected cities in different European regions: Helsinki, Athens, London, Firenze and Gliwice. A part of each city that needed intervention and 'real life' planning alternatives was identified and discussed.

The Community of Practice (CoP) participatory approach that was used to facilitate the interaction between urban planners and BRIDGE scientists, as well as to support the collection of socio-economic data, was described in Chapters 12 and 13. To evaluate specific bio-physical models in BRIDGE case studies, *in situ* (Chapter 5) and remote sensing (Chapter 6) observations were performed. These measurements included standard meteorological data and direct observations of the fluxes of energy, water, carbon dioxide (CO_2) and air pollutants using the Eddy Covariance approach. The major modelling effort in BRIDGE presented several challenges, as described in Chapter 7: the need to downscale the model results to a scale relevant for urban planning; the need to connect models for different environmental components, such as energy, CO_2 and water; the need to respect the constraints in computing time and to ensure the validity of model outputs; and the need to provide comparable reference outputs for the planning alternatives in all case studies in the same time frame. Models ranging from meso-scale air quality to urban canopy models were combined, using a cascade modelling technique from large to local scale to estimate air pollutants (Chapter 8), energy (Chapter 9), water (Chapter 10) and carbon fluxes (Chapter 11). To determine the future distribution of city-wide land uses, a cellular automata model was used to account for the broader effects of planning decisions.

Chapters 14 and 15 presented in detail the Multi-Criteria Analysis (MCA) approach that was used to address the complexity of urban metabolism issues reflected in a wide set of sustainability indicators, most of them developed in the framework of the project. In the framework of BRIDGE, the MCA enabled comparison of planning alternatives, through the structured prioritization of a nested set of variables, concerning specific components of urban metabolism: sustainability objectives, criteria and indicators, defined for energy, water, carbon and air pollutant fluxes, as well as for socio-economic variables relevant for the plans under evaluation. It also enabled the weights and scores for all indicators to be combined into a total assessment index, providing a basis to rank urban planning alternatives.

The DSS that was developed in BRIDGE, based on Geographical Information Systems (GIS), was presented in Chapter 16. More specifically, Chapter 16 described in detail how the DSS can lead the end-user through specific steps to produce: indicators maps for each planning alternative; spider diagrams that show the comparative performance of each planning alternative for each sustainability objective; and a total assessment index for each alternative. The use of the BRIDGE methodology and the DSS to evaluate different planning alternatives in the framework of different future strategic scenarios was discussed in Chapter 17. Finally, Chapter 18, presented guidelines for sustainable urban development that were derived in the framework of the project.

Improving urban planning: the BRIDGE impact

The main impacts of BRIDGE can be divided into three different categories. At the level of impacts on society, it redefined the urban planning exercise to focus on citizens' quality of life through the improvement of environmental quality and reduction of socio-economic costs related to current urban structure and its development. At the level of improved understanding, it brought together different scientific disciplines providing different forms of analysing cross-European variation of physical environments and socio-economic conditions and linking human aspirations and urban design to material and energy fluxes. At the level of information flows, it developed the means of communicating multidisciplinary research findings to urban planning practices and, conversely, communicating real planning problems to research focusing on urban issues, providing therefore the means to close the gap between bio-physical sciences and urban planning and illustrating the advantages of accounting for urban metabolism in planning.

A strategic approach to a sustainable community urban plan involves the redesign of local and regional legislative and political structures and needs to be based on community participation and support. BRIDGE contributed to the formalization of sustainability objectives and provided practical tools for reaching them. Thus, it supported the development of sustainable planning strategies seeking an optimal fit between the urban area and its environment through the creation of a long-term vision, goals and strategies for the allocation of resources and monitoring impacts, as well as detailed action plans. This requires sustained cooperation and exchange of information between different expert groups, which was most effectively accomplished within BRIDGE with the creation of a controlled framework, which forced experts to communicate.

The impact assessment model developed and applied in BRIDGE is based on decision-making theory and impact assessment principles promoting a participative, systematic and coherent approach. The combined integration of environmental, urban metabolism and socio-economic considerations and parameters in the DSS facilitated a comprehensive and holistic approach to sustainability assessment, applicable to varied urban and planning contexts. Furthermore, although it is widely recognized that traditional urban planning models and approaches have contributed to the present environmental situation, it is also clear that addressing environmental issues at city level will not be possible without taking into account climate change. BRIDGE enabled the evaluation of urban planning alternatives in the future scenarios' context by allowing modifications of sustainability objectives and indicators conditional to specific future climate scenarios and then by generating quantitative outcomes for each scenario.

The European Commission's Thematic Strategy on the sustainable use of natural resources highlights the importance of using natural resources in an efficient way, which reduces environmental impacts. The Directive under the Thematic Strategy on the prevention and recycling of waste clarifies the obligation for European Union (EU) Member States to draw up waste prevention programmes at the most appropriate geographical level. The Thematic Strategy on air pollution was developed with the objective of attaining levels of air quality that do not give rise to significant negative impacts on, and risks to, human health and the environment. The urban metabolism approach proposed by BRIDGE is expected to substantially support the implementation of the Thematic Strategies and the related Directives. This is possible because the spatial DSS developed has the potential to interlink the different scales of analysis, from regional to site level, facilitating the comprehension of the underlying mechanisms that drive environmental problems in cities, which in turn can benefit from a site level perspective and thus support the identification of priorities for policy intervention at the city and regional level.

In a framework of sustainable urban planning, actions are needed to adopt an integrated urban metabolism approach and inter-disciplinary research and innovation in developing innovative strategies for waste prevention and management. The DSS can be adapted in order to address aspects not covered within the scope of the project with the main requirement being input of available data. The BRIDGE approach can be extended to highlight how city typologies, drivers and lifestyles can influence urban metabolism, through socio-economic analysis, and highlight the possible benefits to be derived from ecosystems services and green infrastructure. It can be also extended to support science-based decision making and planning for waste management and land use as an integral part of urban development, promoting eco-innovative urban management and re-naturing cities, thus enhancing both the environmental resilience of urban areas and quality of life.

BRIDGE enabled comparisons of the effects of different planning alternatives on physical flows of urban metabolism aspects. The evaluation of the performance of each alternative was done in a participatory way. This interactive process allowed the end-user to gain an understanding of the relative importance of each sustainability objective and indicator. The combined performance and relative importance of indicators were used to rank planning alternatives and the DSS was used to assist the end-users to select objectives and indicators and to define their relative importance. A tool like the BRIDGE DSS may not simplify the urban planning process, but it can help urban planners to deal more adequately with its complexity. Although implementation of the DSS during planning processes may be constrained by lack of resources and skills at municipalities, practitioners can gain significant insight for more informed decision making. The approach could seamlessly be integrated through a proactive attitude towards sustainability and basic upgrading of both planning staff and private sector consultancy skills in municipalities (e.g. GIS and DSS capacity building).

INDEX

T - #0241 - 111024 - C0 - 246/189/12 - PB - 9780367670115 - Gloss Lamination